W0042479

Gösta Kjellsson
Morten Strandberg

Monitoring and surveillance

of genetically modified higher plants

Guidelines for procedures and
analysis of environmental effects

Supported by:

MINISTRY OF ENVIRONMENT AND ENERGY, DENMARK

National Environmental   *NATIONAL FOREST*   ENVIRONMENTAL
Research Institute   *AND NATURE AGENCY*   PROTECTION AGENCY

Springer Basel AG

Editors:

Dr. Gösta Kjellsson
National Environmental Research Institute
Department of Terrestrial Ecology
Vejlsøvej 25
DK-8600 Silkeborg
Denmark

Dr. Morten Tune Strandberg
National Environmental Research Institute
Department of Terrestrial Ecology
Vejlsøvej 25
DK-8600 Silkeborg
Denmark

A CIP catalogue record for this book is available from the Library of Congress, Washington D.C., USA

Deutsche Bibliothek Cataloging-in-Publication Data

Monitoring and surveillance of genetically modified higher plants : guidelines for procedures and analysis of environmental effects / Gösta Kjellsson ; Morten Strandberg. - Basel ; Boston ; Berlin : Birkhäuser, 2001
   ISBN 978-3-7643-6227-0   ISBN 978-3-0348-8317-7 (eBook)
   DOI 10.1007/978-3-0348-8317-7

ISBN 978-3-7643-6227-0

The publisher and editor can give no guarantee for the information on drug dosage and administration contained in this publication. The respective user must check its accuracy by consulting other sources of reference in each individual case.

The use of registered names, trademarks etc. in this publication, even if not identified as such, does not imply that they are exempt from the relevant protective laws and regulations or free for general use.

This work is subject to copyright. All rights are reserved, whether the whole or part of the material is concerned, specifically the rights of translation, reprinting, re-use of illustrations, recitation, broadcasting, reproduction on microfilms or in other ways, and storage in data banks. For any kind of use permission of the copyright owner must be obtained.

© 2001 Springer Basel AG
Originally published by Birkhäuser Verlag in 2001
Printed on acid-free paper produced from chlorine-free pulp. TCF ∞

ISBN 978-3-7643-6227-0

987654321

Contents

Preface

There is an urgent need for guidelines for monitoring of genetically modified higher plants, GMHP. Biotech crops are now cultivated in large scale in North America and elsewhere. In Europe, new genetically modified (GM) products will probably be placed on the market soon and made available for cultivation in the field. Monitoring and surveillance programs for detection of any negative effects to the environment must be designed and ready when these crops are released. This also corresponds to the current intentions made by the European Commission to include monitoring in current biotechnology regulation. Monitoring of changes in biological systems is different from other types of environmental monitoring, such as monitoring fate of chemical pollutants, by focusing primarily on organism survival and organism interactions instead of physical and chemical parameters. The difficulties involved in monitoring biological systems are great, due to the complex interactions between organisms and the variability in responses. Problems concerning spatial and temporal parameter variation increase the difficulties, but may be remedied somewhat by the use of "baselines". These and many other questions are discussed in the present book with the aim of presenting practical solutions to the needs of GMHP monitoring.

A project was initiated in 1998 to produce a book with guidelines for monitoring and surveillance of GMHP. In two earlier books, compilations of current test methods for risk assessment of GMHP were presented (Kjellsson & Simonsen 1994, Kjellsson et al. 1997). The present book extends this work, supplies new facets to the environmental concerns and provides measures for analysis of possible ecological effects when GMHP are cultivated in a large scale. These guidelines for monitoring have been developed in close connection with other ongoing work concerning protocols and detection methods for ecological risk assessment of GMHP (e.g., Kjær et al. 1999).

The project was cofinanced by three organisations in the Danish Ministry of the Environment: the National Environmental Research Institute, the National Forest and Nature Agency and the Environmental Protection Agency.

Members of the project committee Jan Grundtvig Højland, Gitte Silberg Poulsen, Hans Erik Svart, and Helle Nayberg gave valuable comments and recommendations during planning and to earlier versions of the manuscript. Valuable comments and suggestions were also provided by Christian Damgaard, Christian Kjær, Marianne Kruse, Hans Løkke, Vibeke Simonsen and Beate Strandberg.

Valuable technical assistance was provided by Karin FriisVelbæk, Lilian Mex-Jørgensen and Lene Birksø Bødskov. Käthe Møgelvang kindly prepared the drawing for the front cover.

Gösta Kjellsson and Morten Strandberg

Silkeborg, August 2000

1 Introduction

In the present situation, initiatives for monitoring of genetically modified higher plants (GMHP) are planned before we know whether unwanted effects will occur or not. This is contrary to most other types of environmental monitoring, where programs are initiated only after some unwanted effects on the environment have been detected or are found likely to occur.

Risk assessment of GMHP and monitoring as a safeguard

A GMHP-monitoring program will be the follow-up on an intensive risk assessment aimed at detecting and reducing the environmental risks of new GMHP placed on the market for agricultural use. Hence, the likelihood that any adverse effects will occur should be small from the start. It is, however, crucially important, that monitoring procedures and the methods used for data analysis ensure that significant effects will be detected when they do occur. If a monitoring program fails to discover deviations from the baseline or controls this may be a question of statistical power of the tests used for the evaluation of the monitoring results (see Chapter 8).

Risk assessment and monitoring should reduce chances of costly mistakes

Risk assessment and monitoring should act together as an effective control system which will induce the biotech companies to produce the best possible information on possible effects from their products both before and after they are placed on the market. This should be done in order to avoid costly mistakes and additional expenses to both society and industry.

Provision of monitoring data: industry or independent body?

A specific monitoring program should accompany each new case of GMHP and the corresponding risk assessment. The results from monitoring GM plants act both as control of the biotech companies and the risk assessment process. Therefore, it may make sense to have both the design of the monitoring program and the task of data collection placed at a separate institution. Some of the best available knowledge on the GMHP is placed in the industry and in the competent authorities that make the risk assessment. However, while the industry is obviously biased by commercial interests, authorities carrying out the risk assessment may be prejudiced in confirming former assessments. The question remains what alternatives do we have? An independent body or case-by-case committee of research scientists will perhaps be the answer.

Responsibility towards the public interests

Monitoring acts as a safeguard for risk assessment. The current public concern and scepticism about plant biotechnology, which has long been increasing in Europe and now also in USA, warrant that all measures are taken to reduce the risk of the new technology. This makes monitoring an even more crucial and urgent response. Hopefully, the present book will help to fulfil these needs.

1.1 How to use this book

The guidelines described in this book are meant to provide information on the basic concepts and tools available when monitoring of GMHP is planned. The general concepts and definitions are given in Chapter 2 and important conclusions are presented in Chapter 9. Throughout the book, margin text is used to point to central issues for specific subjects. Comments on specific issues where uncertainty prevails are presented in *Italics*. In these cases, we have not been able to recommend a specific method or there may be a lack of generally accepted criteria. A résumé of the content of the chapters are presented below:

Contents - Chapter 1-2

Chapter 1 provides an introduction to the present state in development of GM crops, including major countries and most important plants. The new types of biotech plant products, which can be expected in the near future, are also described. The current version and ongoing revision of EU-directive for genetically modified organisms are discussed. Problems concerning regional differences, limitations and uncertainty in risk assessment are presented with reference to environmental monitoring. In Chapter 2, the concepts of risk assessment and the different types of environmental hazards, which have been suggested for GMHP, are introduced. A list of concepts and definitions is available in Section 2.3 followed by a list of Internet links to the most relevant sites for additional information.

Contents - Chapter 3-4

The objectives for establishment of monitoring programs for GMHP are presented in Chapter 3. The use of three different subprograms is suggested, each covering specific issues, such as monitoring for transgene dispersal, monitoring for effects and procedures for general environmental surveillance. A full scheme for the monitoring process from the definition of objectives, data collection, analysis, evaluation and decision making is presented. Experiences of monitoring from different fields of research and the specific factors, which influence the results, are discussed. In Chapter 4, the problems concerning selection of relevant habitats for monitoring and surveillance of GMHP are discussed. Sections on general changes affecting plants and animals and baseline determination are included.

Contents - Chapter 5-6

Environmental issues concerning the use of GMHP in farming and detection of effects from small-scale field trials and large-scale releases are discussed in Chapter 5. Monitoring and detection of GMHP dispersal and invasion are treated in Chapter 6. Problems concerning crop-wild species hybridisation and the use of markers for detection are discussed.

Contents - Chapter 7-8   Chapter 7 considers the different types of possible effects from GMHP on organisms and ecosystem in detail and presents suggestions for detection. Specific monitoring suggestions are given for soil animals, insect pollinators and herbivores, mammals and birds. Different statistical procedures, which are useful when monitoring is planned, are presented in Chapter 8. These include power and trend analysis and sensitivity analysis with relevant examples of the use given.

Contents - Chapter 9-11   Important conclusions and recommendations for monitoring of GMHP are presented in Chapter 9. This involves both general procedures and specific matters concerning detection and analysis. Chapter 10 contains the list of references for the book made from a database in the bibliographic system EndNote. The index, Chapter 11, at the end of the book will point to specific references of concepts, common and Latin species names.

1.2 Biotechnology and new GM crops placed on the market

GM crops grown in 1999   The use of biotechnology in plant agriculture is becoming increasingly important worldwide. In 1999, the transgenic crops which were cultivated over the largest areas were soybean (21.6 mill. ha.), maize (11.1 mill. ha.), cotton (3.7 mill. ha.) and oilseed rape (3.4 mill. ha.) (James 1999). The most important countries growing GM crops were in descending order: USA, Argentina, Canada, China, Australia and South Africa. USA alone contributed with 72% of the global area (28.7 mill. ha.). Herbicide tolerance and insect resistance (Bt toxins) were totally dominating as GM products for the global market (Table 1.1).

Trends in commercial GM   There was a general three-fold increase in herbicide tolerant GM crops
plant development   from 1997 to 1999. In the same period, the use of crops with resistance to insect attack only, showed a small 1.2 fold increase, while GM crops with combined Bt and herbicide tolerance increased strongly by a 69.7 fold. This also illustrates the tendency seen from biotech research and field releases that the number of different traits inserted into the same plant tend to increase.

Based on the field trials applications and research literature, the current development is from the first generation of single gene traits, such as herbicide tolerance, insect and fungal resistance, towards a second generation of GM plants with multiple gene inserts covering new plant properties (Dunwell 1999). The new development is most evident in USA and Canada, but also present in field trials in Europe. The public and environmental concerns towards new biotech products, which lately has been increasing in Europe, will clearly influence the speed of this development.

Table 1.1 Global area of dominant GM crops and traits in 1999 and the increase from 1997. Source: James 1997, James 1999.

Trait / Crop	Million hectares	Increase ratio
Herbicide tolerance	28.1	3.1
soybean	21.6	3.2
canola/rape	3.5	1.9
cotton	1.6	3.0
corn/maize	1.5	6.5
Insect resistance (Bt)	8.9	1.2
corn/maize	7.5	1.5
cotton	1.3	0.2
Bt/Herbicide tolerance	2.9	69.7
corn/maize	2.1	-
cotton	0.8	-
Virus resistance/ Other	<0.1	-

Second generation of GM plants and traits

The major new type of GM plant traits and products we can expect, include:

- Increased stress tolerance (e.g., drought, frost and salt).
- Changed growth characteristics (e.g., male sterility system, fruit ripening phenology, increased growth rate and nitrogen fixation).
- Changed chemical composition (e.g., proteins, oil content, starch and vitamins).
- Production of pharmaceuticals (e.g., vaccines, hormones and enzymes).
- Production of industrial compounds (e.g., biomass, lignin and plastics).
- GMHP with inducible promoters that activate specific traits (e.g., pest resistance and flowering) when needed.
- Bioremediation (decontamination of toxic soils).

The major crops used worldwide will still be the most important objects for modification, but new functional types of plants, such as fodder grasses, fruit-trees and trees used for wood production (e.g., *Populus*, *Abies* and *Eucalyptus*), may become increasingly important. The new trends in biotech development has significant consequences for both risk assessment and monitoring in terms of possible new adverse effects and the way these environmental changes are detected.

1.3 Risk assessment, monitoring and the EU directives

EU directives for GMHP and GM organisms

The various issues of risk assessment of GM organisms (GMO) including higher plants (GMHP) are thoroughly treated in EU regulations including the directive 90/220/EEC on "the deliberate release into the environment of genetically modified organisms", which is now under revision. A new Annex VII to the directive is relevant for monitoring of both GMO encompassing GMHP and will include terms and objectives for a monitoring plan and provide some guidance for how this shall be carried out. A positive step in the new amendment is that risk assessment will include not only direct and immediate effects but also indirect and delayed (i.e., long-term) effects in assessment and monitoring.

Risk assessment procedures in EU

The environmental concerns are treated in EU risk assessment procedures that each particular case go through before it is decided whether the product can be placed on the market (Directive 90/220/EEC). The basic working tool is the "Summary Notification and Information Format" (SNIF). Extensive information, provided by the applicant biotechnology company, is circulated in member countries in each case. Technically, an application is made for one country, and if the assessment is positive, other member countries can comment and make requests for additional information. Permission for placing on the market is valid in all EU countries, so the GM crop can be cultivated almost anywhere in Europe.

Lack of EU test-procedures for GMHP

Although the EU-directive stipulates the type of information required for risk assessment, it gives no specific guidance to how this should be provided. Test guidelines, which follow a tiered approach, similar to those generally used in environmental risk assessment, have been suggested also for GMHP (Kjellsson 1997, Strandberg et al. 1998, Kjær et al. 1999). Hopefully, it will soon be decided to include some sort of structured test procedures in the EU for GM plants.

Regional differences with focus on NW Europe

The specific objectives for monitoring will vary in different regions of Europe. Depending on climate, wild flora and fauna and cultivation practices, the particular hazards caused by, e.g., hybridisation, interactions with the environment and crop use, differ somewhat between regions. The suggestions made in the present work apply primarily to NW Europe, covering the temperate climatic zone of Denmark, Sweden, Norway, Finland and parts of Great Britain, the Netherlands and Germany.

Limitations and uncertainty in risk assessment

The environmental risk assessment of GMHP made by the official authorities will ideally point out any major threats to the environment. However, even a carefully made assessment will always depend on the level of information and scientific knowledge available at the particular moment. These is also uncertainty in research results caused by statistical, spatial and year-to-year variability, which always exist (see Section 4.6). Furthermore, the complexity of natural ecosystems makes predictions based on simple data, such as agricultural field tests, difficult to perform. Thus, the limitations of the risk assessment partly depend on the unpredictability of the environment for release (see above), on uncertainty factors and on the specific GMHP and its interactions with other organisms and the environment.

Predicting effects of GMHP are difficult - therefore monitoring is needed

As we are unable to predict if any unexpected effects will arise from the release of a GMHP, we need to look for changes in the environment. These changes may be caused by many different factors besides the GMHP itself, which will require careful analysis and research (see Chapter 4). Furthermore, known effects which were considered insignificant at the time of assessment, may perhaps later become a problem. This has been an experience from the research done on plant invasion, where there generally is a time lag of perhaps 5-50 years between the first introduction and the time it becomes a problem (Hobbs & Humphries 1995, Marvier et al. 1999). The need for long-term monitoring has been stressed, as predicting the ecological behaviour of a species in a new environment may be effectively impossible (Blossey 1999, Williamson 1999). Consequently, there is a need for monitoring to confirm that the assumptions of the risk assessment are still correct and to detect any unexpected effects that may arise.

Monitoring and risk management

The information collected by a monitoring program for GMHP should be important not only to risk assessment but also to the management of any adverse environmental effects which may have been detected during the program. It is important that detection of any unwanted change lead to prompt and proactive responses in risk management measures if these are going to be successful. However, if indirect effects to, e.g., organisms in food chains, etc. are suspected, knowledge based on research and analysis will be required before the relevant measures can be taken. The majority of measures will probably concern reduction of the detected adverse effects, e.g., through changes in farming practices, weed management or, in worst case, a cease of the marketing permission. Some processes, such as transgene dispersal and hybridisation, are irreversible in nature. If a GM hybrid cause problems this may lead to a situation with the need of permanent management in affected ecosystems.

2 Environmental concerns and concepts

Right from the beginning of the development of plant biotechnology serious concerns for the possible adverse effects to natural environments were raised (see, e.g., Tiedje et al. 1989). These concerns focus on the probability that the GMHP will cause an effect with negative consequences to the environment (i.e., hazard), often expressed as (UNEP 1996):

$$Risk = Probability \times Hazard$$

Objectives in risk assessment of GMHP

In the process of ecological risk assessment, the likelihood of occurrence and the severity of the effects become central issues. The prerequisites for any effects to occur are, of cause, that the GMHP or the genetic construct (i.e., the transgene) is present in affected habitats or in the surroundings. Consequently, issues concerning GMHP invasion and hybridisation with wild relatives become primary issues to ecological risk assessment and to monitoring when modified crop plants are used in field trials or commercially released. The effects that released GMHP with altered traits may have to the environment are theoretically numerous, but until recently no significant effects had been demonstrated even in North America. Lately, however, indications from mainly laboratory trials suggest that adverse effects may occur to non-target insects from the use of GM plants with Bt-toxins or other types of transgenic insect-resistance (Hilbeck et al. 1998, Jørgensen & Lövei 1999, Losey et al. 1999, Hilbeck et al. 2000).

2.1 The risk of GMHP invasion

The risk that genetically modified plants may invade natural habitats and cause environmental problems have been put forward since the start of biotech farming (Regal 1993, Williamson 1993). This view has largely been based on experience from invasions of exotic plants. However, the majority of important crop plants, such as maize, soybean, cotton and potato, do not in general run wild under conventional farming in Europe or North America.

Invasive crops and cultigens running wild

Some crops, such as oil-seed rape, may occasionally invade field surroundings, roadsides and wasteland; but rarely progress into semi-natural habitats (Crawley & Brown 1995). Hitherto, there has been very little direct evidence of increased invasion caused by GMHP. An extensive use of GM plants with increased tolerance to environmental stress (e.g., drought, salt or cold) or changed life-cycle components could increase the risk of invasion. Furthermore, when wild relatives of the GMHP exist nearby the fields, and hybridisation is possible, the situation becomes more critical. Some crops, such as alfalfa

(*Medicago sativa*) and carrot (*Daucus carota*), which have closely related wild species in natural habitats, may be in a particular high-risk category (Jacot & Ammann 1999). The list of cultigens running wild in Central Europe is extensive and besides those already mentioned, includes (Bartsch et al. 1993): grasses (e.g., *Festuca pratensis*, *Lolium perenne* and *Setaria italica*), ornamental plants (e.g., *Narcissus spec.*, *Impatiens glandulifera* and *Reynoutria japonica*) and trees (e.g., *Prunus serotina* and *Acer* spp.).

Gene-flow into GMHP from field surroundings may require a survey

Weedy genes may introgress into the GMHP field crop from areas surrounding the field. This can potentially change life-cycle components (e.g., seed dormancy, competitive interactions) of the crop so that it becomes more persistent and weedy. While this is primarily a problem to practical farming and, e.g., herbicide management, it may also result in hybrids, which are more invasive in the field surroundings. An example for comparison is *Beta vulgaris*, where wild beets (ssp. *maritima*) are able to introgress with sugarbeets (ssp. *vulgaris*). The hybrid "weed beet", which is able to bolt in a single season, has become a serious problem to farming in Northern France, Belgium and Germany (Parker & Bartsch 1996). A survey on the occurrence of wild relatives to a GMHP in exposed habitats close to the field will give valuable information on the risk of introgression. See Chapter 6 for more information on gene flow.

2.2 Ecological effects of GMHP

An extensive range of potential hazards to the ecosystem caused by GMHP has been suggested in literature (see, e.g., Colwell 1994, Crawley 1995). The main types of ecological hazards may be listed as follows (see also Kjellsson 1997, Strandberg et al. 1998):

- Direct invasion of natural ecosystems (see above).
- Gene transfer and hybridisation with wild relatives (see above).
- Disruption of natural communities through competition or interference.
- Harm to non-target species, e.g., by indirect effects in food-webs.
- Loss of biodiversity or genetic diversity.
- Changes in nutrient cycles, primary production and geochemical processes.

The first two points in the list are strictly speaking not effects but prerequisites for the actual hazards to occur.

Great complexity between organisms and effects

The main types of organism groups and different biological processes and interactions involved in potential changes to the environment are listed in Table 2.1. The complexity of potential organism interactions and effects is evident.

Table 2.1 Organism groups and biological processes relevant for monitoring of environmental effects of GMHP. Both direct and indirect effects may be monitored on different combinations of organisms and processes.

Organism group/organism	Processes and interactions
Pathogens (vira and fungi)	Population and organism processes
	Changes in genetic diversity
Plants	Population size changes
GMHP	Life cycle changes
hybrids, weeds	
	Organism interactions
Soil detrivores	Altered foodweb interactions
Collemboles	Altered mutualistic interactions
Earthworms	Altered competitive interactions
Insects	Ecosystem processes
pollinators	Vegetation changes
herbivores, predators	Changes in primary production
	Changes in nutrient cycles
Birds	Changes in geochemical processes
Mammals	

The main types of adverse effects from GMHP to the cultivated ecosystems are (Snow & Palma 1997, EEA 1999):

- Adverse changes to farmland flora and fauna through continued dependency of monocultures and pesticides.

- Cumulative environmental impacts from multiple releases and interactions.

- Increased problems with pests and weeds through evolution of resistance to herbicides and build up of pest and disease resistance.

Hazards to sustained cultivation

In the farmland, extensive use of GM crops presents special problems to growers of conventional or organic crops by pollen dispersal, hybrid formation and volunteers (Moyes & Dale 1999, see also Section 5.2). Widespread cultivation of genetically modified crops could also speed up the evolution of undesirable weeds and pesticide resistant insects (Tiedje et al. 1989, Rissler & Mellon 1993). These types of hazards are not strictly relevant to an ecological risk assessment but mainly affect the possibilities of a sustained cultivation of a particular modified crop type. Some sort of management procedure, such as the use of refuges, is commonly applied to reduce the risk of resistance building up in target organisms.

2.3 Concepts and definitions

In this section, the definitions of important concepts and issues employed in the present manual and relevant to monitoring are presented. The usage is generally in accordance with terminology and procedures within the EU, and any differences reflect the personal views of the authors.

The list is presented in alphabetic order and references to corresponding terms are given. The meaning of some important acronyms has been included in the list.

Concept	**Definition**
Acceptable effect	An effect of the GMHP on the environment, which occur at a low level and is assessed as being ecologically insignificant and not adverse. This term has not been implemented in current regulation. See also *unacceptable effect* and *precautionary principle*.
BACI	Before-After, Control-Impact; principle used in planning and when carrying out effective environmental monitoring.
Baseline information	The base or expected normal situation including variation from which future change in a habitat may be detected (see Section 4.6).
Basic state	The composition of biotic (e.g., species and populations) and abiotic variables in an ecosystem just before or when monitoring starts. See also *baseline information*.
Biodiversity	Number and relative composition of species or ecotypes in a habitat.
Bt	*Bacillus thuringiensis*, a bacterium which can produce compounds which are toxic to insects. Bt-toxin has been biotechnologically introduced in GM crop plants for protection (resistance) from attack of insect larvae or adults.
Case-by-case	Both risk assessment and monitoring of a GMHP are done separately for each new case.
Cultivar	A variety of a crop species grown for agricultural production.
Delayed effect	Any effect on the environment which is observed after the period of the release of the GMHP has terminated or at a later stage (suggestion: In this context, 5 years or more after release). See also *immediate effect*.
Decision-making	Process of arriving at decisions, especially in organisations or in government. Decision-making may involve complex issues which are discussed in large fora.

Direct effect	Any effect on the environment which is the result of the GMHP itself. Effects that occur through a causal chain of events are not included. See also *indirect effect*.
Disturbance	Inflicted physical changes to vegetation, organisms or other components in the ecosystem. Includes man-made disturbances (e.g., soil scarification, clearing, burning etc.) and natural disturbances (e.g., fire, storms, perturbations caused by animals).
Early detection	When critical changes in environmental parameters are detected early, management measures will be more effective and less expensive than later.
Early warning	Adverse effects which are detected at an early phase of GMHP use or dispersal, may point to problems which will become serious at a later stage. See also *early detection*.
EU	European Union.
Functional diversity	Diversity of different functional groups in a community of species. See also *genetic diversity* and *functional group*.
Functional group	A heterogeneous array of species, where each members of the same functional group has a similar role in the functioning of the ecosystem. A community of bacteria can, e.g., contain the functional groups: nitrifiers, ammonifiers, denitrifiers and others.
Genetic diversity	Amount of genetic polymorphism which is present within the genomes of all the individuals of a population or a species.
GM	Genetically modified, e.g., a crop, a plant species or any other type of genetically modified organism.
GMHP	Genetically modified higher plant.
GMO	Genetically modified organism.
Hazard	An undesired (adverse) effect to the environment.
Hazard identification	In risk assessment and in monitoring the different potential hazards caused by the GMHP must be identified. Monitoring must be specifically targeted at risks, which have been previously identified.
Immediate effect	Any effect on the environment that is observed during the period of the release of the GMHP. See also *delayed effect*.
Indicator species	Species or assemblages of species used in assessing environmental change in habitats.

Indirect effect	Any effect on the environment which occur through a causal chain of events as the result of interactions of the GMHP with other organisms, transfer of genetic material to other organism or changes in use or management. See also *direct effect*.
Large-scale release	The release of a GMHP in a large area or in several places, typically when the GM crop has been placed on the market as a commercial product. See also *small-scale release*.
Long-term (effect)	Effects on the environment that occur during a long period of time after the initial ecosystem perturbation, e.g., the GMHP release. See also *short-term (effect)*.
Marker	A detectable trait for a genotype, found in an individual plant. Main types of markers include: morphological, tolerance (e.g., to a herbicide or an antibiotic), proteins (e.g., isozymes), antibodies and DNA profiles (e.g., the genetic construct).
Monitoring	Data sampling and detection of environmental changes in predefined parameters (e.g., organisms, processes). The purpose for monitoring is related to a specific cause, here the introduction of a GMHP. See also *surveillance*.
Non-target organism	Organism that is unintentionally affected by the biotechnologically inserted trait. See also *target organism*.
OECD	Organisation for Economic Co-operation and Development.
Parental organism	The particular organism from which the transgene insert originates.
Policy	Principle which is adopted or pursued by an individual, an organisation, government or industry.
Precautionary principle	A rule that allows governments and competent authorities to impose restrictions and initiate actions towards possible environmental problems. The principle emanates from the wish to protect man and nature even if there is insufficient scientific evidence of the extent and cause of the environmental damage.
Promoter	Region of DNA involved in gene transcription and regulation of protein formation by the gene.
Recipient organism/ plant	The particular organism/ plant species, subspecies or cultivar which receives the transgene insert.
Refuge	An area with susceptible plants near an insect resistant or herbicide tolerant crop. Refuges are commonly used to slow down the positive selection of Bt-resistant insect populations and occasionally to deter development of herbicide-tolerant weeds.

Risk management	All measures taken to prevent or minimise potentially adverse effects from deliberate release or post marketing of a GMHP.
Short-term (effect)	Effects on the environment that occur within the first 3-5 years after the initial ecosystem perturbation, e.g., the GMHP release. See also *long-term (effect)* and *direct effect*.
Small-scale release	The release of a GMHP in one or a few places covering small areas, typically for experimental purposes. See also *large-scale release*.
Statistical power	The probability of getting a statistically significant result when monitoring an existing effect. See also Section 8.1.
Surveillance	General observations and data sampling for detection of non-specific environmental changes. See also *monitoring*.
Target organism	Organism that is primarily and intentionally affected by the biotechnological inserted trait (e.g., an insect species susceptible to a specific Bt-toxin or a pathogen). See also *non-target organism*.
Transgene	The gene construct that determines a biotechnological inserted trait.
Transgene stacking	Deliberate introduction of multiple transgenes with different traits into the same plant. May also occur through unintentional gene flow from crops to other crop cultivars or to weeds.
Trend detection	Measures of the amount and level of change in environmental parameters during a period of time (parameters include species abundance, population growth, etc.).
Type I error	Statistical decision error, the probability of rejecting the null hypothesis when it is true. See also Table 8.1.
Type II error	Statistical decision error, the probability of accepting the null hypothesis when it is false. See also Table 8.1.
Unacceptable effect	An effect of the GMHP on the environment which occurs at a high level and is assessed as being ecologically significant and adverse. This term has not been implemented in current regulation. See also *acceptable effect*.
Volunteer	Crop plants which propagate from last years crop and sustain in the field or the immediate surroundings.
WTO	World Trade Organisation.

2.4 List of relevant links for information on monitoring of GMHP

A selection of Internet links is included with useful information on releases of crop species, modified traits, biology and ecology of organisms and general information on environmental monitoring. Although the links have been checked just before publishing, addresses may become obsolete at short notice.

2.4.1 Links to GM crops, wild plant species and field tests

Agricultural Research Service services the database GRIN with information for cultivated plant species on taxonomy, distribution world wide, and lists of cultivars used in single countries.
http://www.ars-grin.gov/npgs/tax/index.html

GM plants and farming in UK

An introduction to issues and concerns of GM plants are available in a link to John Innes Centre, UK. A broad range of aspects is treated, including: ethics, consumers choice and effects to organic farming, besides the more technical issues.
http://www.gmissues.org

NCBE on GM food and crops in UK

The National Centre for Biotechnology Education (NCBE) in UK provides a broad overview of GM crops and food regulation with links to major reports and important research and information at: http:// www.ncbe.reading.ac.uk

EU field trials

Information of deliberate fields trials of GMHP under the directive 90/220/EEC within the European Community are best assessed from: http://food.jrc.it/gmo

Links to the sources for international field tests, currently covering 22 countries at: http://www.nbiap.vt.edu/cfdocs/globalfieldtests.cfm

OECD databases and publications

OECD's database on field trials "Bio Track" offers reviews of GMHP trials worldwide, but is often not as up to date as national links on the subject. This valuable summary information worldwide is available at: http://www.olis.oecd.org/biotrack.nsf

OECD has a range of "publications on harmonization in biotechnology" of which the consensus documents with biology of GM plant species are especially valuable: http://www.oecd.org/ehs/public.htm

USDA permits and databases

USDA, the US Department of Agriculture, Animal and Plant Health Inspection Service, provides information on biotechnology permits at: http://www.aphis.usda.gov/biotech/

USDA provides access to a database containing standardised information on both cultivated and wild plant species. Up to now it mostly provides data on taxonomy, distribution in the US and lists of taxonomic references. http://plants.usda.gov/plants/

GMAC release information from Australia

GMAC, the Genetic Manipulation Advisory Committee, provides information, with annual case summaries, and guidelines for releases of GM plants in Australia.
http://www.health.gov.au/tga/gene/gmac/gmac.htm

2.4.2 General links to monitoring

Biosafety bibliographic database

Biosafety is a scientific bibliographic database on "Biosafety and Risk Assessment in Biotechnology". The database is updated monthly and contains, to date, about 2000 scientific articles (full reference + abstract), published on international scientific journals since 1990, selected and classified by ICGEB scientists for the main topics of concern for the environmental release of GMO.
Available at: http://www.icgeb.trieste.it/~bsafesrv/

EMAN information from Canada

EMAN, Ecological Monitoring and Assessment Network, gathers information on research and monitoring of plants and animals in Canada. Available information include, e.g.: monitoring protocols, frame programs, networks and the effects on biodiversity from a range of environmental factors, including GMHP.
http://www.cciw.ca/eman-temp/intro.html

IHGE, the Information Highway to the Global Environment, presents a wide overview on the organisations, which are concerned with environmental monitoring in different countries.
http://www.gsf.de/UNEP/contents.html

UNESCO's system for classification of natural vegetation is available in a modified version from USGS National GAP Analysis Program.
http://www.gap.uidaho.edu/gap/AboutGAP/Handbook/Misc/UNESCO/unesco.htm

USGS monitoring information

USGS monitoring program determines the state of and trends for animal and plant life in the USA and develops prognostic tools for prediction of future changes. Furthermore, the program tries to identify populations, species and ecosystems which are threatened. It also provides a simple online power-analysis program for determination of the number of samples needed and information on natural variation in plant and animal populations. http://www.mp1-pwrc.usgs.gov/

World Conservation Monitoring Centre gives information services on conservation, international conventions and research programs. This includes vulnerable plant species (incl. trees) and red lists for threatened animals.
http://www.wcmc.org.uk/species/data/index.html

2.4.3 Links to monitoring analysis tools

A GAP-program for analysis of fragmentation and detection of changes in the landscape structure is available on: http://www.gap.uidaho.edu/gap/Tools/Index.htm

The free-ware program "Monitor" for statistical power analysis of monitoring programs is available for downloading on: http://www.mp1-pwrc.usgs.gov/powcase/monitor.html

The State University of Iowa provides information on power analysis and the needed number of samples for some common types of tests. The program "PiFace" is made available for determination of power values on: http://www.stat.uiowa.edu/~rlenth/Power/

3 Monitoring and surveillance of GMHP dispersal and environmental effects

In this chapter, the regulatory background and the objectives for monitoring GMHP are presented in the context of the EU directive. From this base, a monitoring scheme consisting of three subprograms is suggested. A total monitoring scheme is presented, involving different steps from definition of objectives to decision making within a cyclic approach. Different methods and suggestions are given both for monitoring in general and for the subprograms in specific.

3.1 Objectives for monitoring programs

EU purpose of monitoring and survey of GMHP

The text in the annex VII to the EU directive 90/220/EEC on deliberate release of GMHP into the environment, contain the following objectives for a monitoring plan:

- "confirm that any assumption regarding the occurrence and impact of potential adverse effects of the GMO or its use in the e.r.a. (i.e., the environmental risk assessment) are correct, and

- identify the occurrence of adverse effects of the GMO or its use on human health or the environment which were not anticipated in the e.r.a.".

The intentions are that monitoring targeted at transgene dispersal and specific effects is done covering 1 (case-by-case), while a non-specific survey for unforeseen effects is done covering 2. The administrative need for confirmation of the assumptions, and consequently also the decisions, made in the assessment process is well founded. It should be noted that environmental problems caused by, e.g., cultivation and management practices have been included. The detection of unforeseen adverse effects by a general survey is more problematic to implement in practice (see below and Chapter 7).

General objectives for monitoring and survey of GMHP

The different environmental concerns described in Chapter 2 and the intentions in the new EU Annex VII suggests that the general objectives for monitoring and surveys can be stated as follows:

1. a. Detection of GMHP invasion and hybridisation and transgene dispersal. Monitoring procedures are dependent on species case and based on risk assessment information.

    b. Detection of adverse impacts caused by the GMHP.
        - Environmental effects in natural habitats where the transgene or the GMHP is present.

- Environmental effects caused by the use of the GMHP in the field or in field surroundings (herbicides, single crop systems, etc.).

Monitoring procedures depend on inserted trait and plant species (life-form, etc.).

2. Detection of unforeseen adverse effects in natural habitats, in the field and field surroundings. Survey of any major effects to ecosystem components. The survey is not case or species specific.

Three subprograms

Three subprograms are presented which aim at different problems that arise in connection with monitoring GMHP and their effects. Thereby the three subprograms make up a unity that hopefully will be able to detect both foreseen unlikely effects and unforeseen effects.

1. Monitoring of transgene dispersal

The first subprogram is designed solely to detect dispersal of the GMHP itself or the inserted transgene. This can happen by spread of the seeds in space or time (i.e., through the seed bank), vegetative fragmentation, and by pollen dispersal and hybridisation with non-modified populations of the same species or with related species. In practical terms, this type of monitoring is relatively easy to design and to sample relevant data for. The occurrence of the GMHP or the transgene in the environment is often a prerequisite for the occurrence of adverse effects to other organisms or to the ecosystem. If adverse ecosystem effects from dispersal are detected, it should lead to reassessment of the GMHP.

2. Monitoring for effects

The second subprogram is designed to detect adverse effects of the GMHP or its use on the surrounding ecosystem in form of altered ecosystem function or composition. This part of the monitoring program is suggested located on governmental areas or areas purchased or rented by the GMHP manufacturers. These areas are to be monitored permanently for a long-term period with large-scale investigations in permanent plots laid out at different distances from the fields where GMHP are grown. There will be difficulties in explaining any effects indicated by the monitoring data as direct or indirect results of the GMHP. Consequently, this program should perhaps only be implemented if actual transgene dispersal or GMHP invasion into the area has been detected by the first subprogram.

3. Surveillance

The third major issue is the surveillance subprogram that provides an overview of the GMHP on a wider but less intensive scale. This part of the program should be carried out by the GMHP-farmers, their consultants and volunteers. Instruction concerning surveillance should be distributed to farmers that want to grow GMHP crops.

Early warning

One major purpose of monitoring and surveillance of GMHP is an early detection of any adverse effects to the environment caused by the

GMHP or its use. An early warning provided by information on major changes to the field ecosystem or the immediate surroundings will, at least in theory, allow that management measures are taken to reduce the negative impacts.

Need of criteria for adverse effects

An important prerequisite for a monitoring program is information about criteria of unacceptable adverse effects. Such information is lacking, partly because there is no experience of what to expect, partly because such acceptability limits have not yet been decided politically. In the absence of criteria, the aim must be to suggest a monitoring program that measures effects at such a level that the program with a reasonable effort can be implemented. This also means that the program may not detect some minor changes that are caused by a particular GMHP or its use. However, the results from monitoring may suggest need of further trials and ideas for extension that may be able to detect smaller effects. Furthermore, the surveillance program is intended to detect major unforeseen effects, which are not covered by monitoring. In these cases, the precautionary principle in risk assessment may lead to the decision that the GMHP is withdrawn from the market.

3.2 Scheme for the process of monitoring GMHP

A proposal for a general scheme for monitoring GMHP with separate steps for different aspects is shown in Figure 3.1. The monitoring process starts with information on the specific case (centre of Figure) and proceeds through definition of objectives, selection of methods and design, monitoring with data collection, analysis of data, evaluation and decision making. The process shown typically covers a period of minimum one year, but an important aspect is that it is repeated in cycles for a period of years. This is of cause needed for acquisition of time series data, but also essential to flexible decision making, reassessment of specific cases and adjustment of monitoring procedures. The different steps in the monitoring process is shortly described and discussed below. Steps overlap to a certain degree and flexibility in the process is obviously needed to make an effective monitoring program.

Definition of objectives

The objectives of the monitoring program have to be defined and so clearly stated as possible. Some aspects of monitoring GMHP are general (e.g., dispersal and hybridisation) while others are strictly case- and trait-specific. In the definition process, any potential adverse effects identified in the assessment, but considered not significant for rejection of release, must be addressed. Monitoring GMHP fate and effects should in fact become a direct extension of the assessment procedure. For general issues see Table 2.1. Particular issues on dispersal and hybridisation and on effects are described in Section 6.1 and 6.2, respectively (see also Kjær et al. 1999). Ideally, baseline information, e.g., on composition and diversity, should be used already at this stage to assess the possibilities of getting the relevant data within a reasonable time scale.

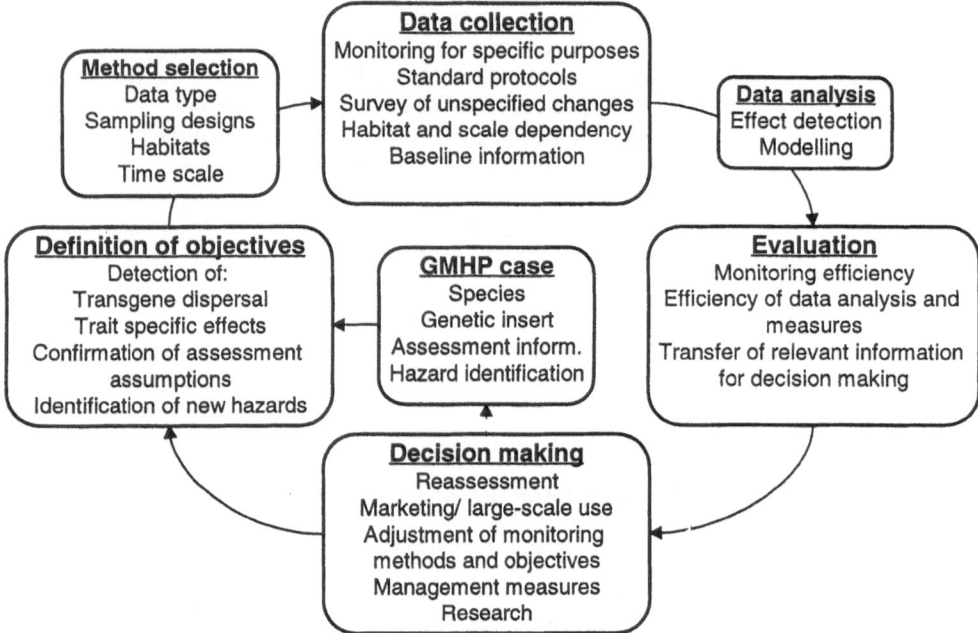

Figure 3.1 Scheme for a process for monitoring GMHP invasion and environmental effects. The process begins with information on the GMHP in the centre and proceeds to definition of objectives for monitoring to the left. A cyclic process is followed and repeated at regular time intervals. For details on process steps see text.

Method selection

Background information on the particular crop species, inserted trait, cultivation practices, etc., should be used in planning and design of the monitoring plan. Sampling and plot designs need to be carefully made based both on particular issues of concern (see Chapter 6 and 7) and statistical considerations, e.g., power of tests (see Chapter 8). The types of habitats involved (e.g., semi-natural or cultivated areas, see Chapter 4 and 5) and the time scale also need particular attention. Sampling design should provide guidelines for the most cost-efficient way to collect data with high quality standards. The uses of the collected data for decision making and reassessment of GMHP release permit should be clear.

Data collection (monitoring)

Trained personnel are required for data collection for specific purposes, especially when biodiversity, particular organism groups and gene transfer are concerned. Standard protocols should be used to make the quality of data consistent. The timing in relation to plant phenology and farming practices must be considered. Farmers or agronomic consultants can make surveys of major unspecified changes and volunteer escape in farmland areas and field surroundings if clear instructions are provided. Standard formats for data in databases or spreadsheets must be used. Direct logging or registration of data in portable computers is to be preferred.

Data analysis

The type of statistical data analysis will already have to be decided at the planning stage of monitoring. The analysis methods should then by now be ready to be used in a standardised way to detect significant effects or trends in GMHP populations (e.g., increases) or populations of non-target organisms (e.g., decreases). A quality check of the data-transfer from paper-notes to computer or from logger to computer is important. Detailed data documentation must be made. Results must clearly relate to the stated objectives for the monitoring program. The results from the data analysis should be in form of a synthesis or report which can be used by field researchers, managers and decision-makers.

Evaluation

An overall evaluation of the results of the monitoring program includes the efficiency of measured data and the analysis in providing answers for the questions initially asked. Evaluation thus involves the efficiency of the sampling methods and the relevance on time spent on particular objectives. At this point, correction of sampling procedures and statistical methods may be needed. If specific models are used for predictive purposes, a validation based on collected data and necessary adjustments can be made. The results made available to decision-makers should be clearly stated as far as possible in context of previously identified or any new identified risks caused by the use of the GMHP.

Decision-making

Decisions based on the results from monitoring and surveys will involve a possible reassessment of the permit for release or for commercial use and placing on the market. New hazards may also have been identified, which can lead to new monitoring objectives that must be included. In the light of findings, adjustment of methods and monitoring goals may be necessary. Appropriate responses to temporary results will have to be made also concerning the onset of management measures to reduce negative impacts on vulnerable habitats and organism groups (i.e., risk management).

Quality loop of monitoring

The cyclic nature of the monitoring process, with continuous quality improvements from year to year, has been called the quality loop or quality spiral (Ekedahl 1997). It is stressed that there has to be a balance between the need for improvements and the necessary stability and progress of the monitoring process.

3.3 Approaches to monitoring

Design-based versus model-based monitoring

Two major approaches are generally applied to monitoring of environmental change: a traditional design-based inference versus a model-based inference (Edwards 1998). Design-based approaches requires probability sampling and relatively rigid design; produces objective results within the frame of assumptions, but may be costly and lack flexibility. Model-based approaches are usually flexible and less dependent on rigid sampling, but major assumptions, which need to be made, induce bias which may be hard to estimate. Furthermore, models must be validated in the ecological context.

Design-based monitoring of GMHP

In the present work we primarily suggest the use of standard, design-based procedures for monitoring of GMHP invasion and effect detection. However, the use of modelling is commented on in Section 8.2 and some references given for the interested reader.

BACI design for monitoring environmental effects

The Before-After, Control-Impact (BACI) design has been much used as a basic concept for monitoring and detection of environmental disturbances (Underwood 1994, Philippi et al. 1998). It reflects the importance of having relevant data from before an impact was started. Otherwise changes may be difficult to quantify and test statistically. Controls are especially needed for comparisons of effects in a naturally changing environment (e.g., species turnover, succession of vegetation and animal groups in habitats). In practice the "Before data" may be confounded by fluctuations in population sizes etc., so knowledge of temporal variation becomes essential. In the traditional BACI design only one control site is used. If the number of controls is increased to equal the number of test areas, data on temporal fluctuations between sites can be used in detection of significant impacts (Underwood 1994).

BACI and GMHP monitoring

When monitoring for effects of GMHP use, general data on organism groups (weeds, pest insects, birds, crops and districts, etc.) may be available before impact start. However, data from individual test fields and surroundings will provide a better base for detection if monitoring is possible. Relevant controls may be hard to find, but field and habitats of similar biotic composition and soil type without immediate GMHP exposure would be a first choice (see also Section 4.6 on baseline definition, Section 7.5 on partial replacement, and Chapter 8). The use of controls and application of the BACI principle in cultivated fields is described in Section 5.1.

3.3.1 Quality of monitoring programs

Factors which influence the quality of monitoring programs

Numerous factors may influence the quality of monitoring programs (Bakke et al. 1997, Ekedahl 1997, Johnsen 1997). The major factors of importance are:

- Representative sites.
- Within site variation.
- Temporal variation.
- Methodological uncertainty.
- Taxonomic problems.
- Statistical considerations.
- Personnel and equipment.

Representative sites

The selected sites should without being too numerous represent the typical conditions of the region where a certain GMHP is grown. These conditions should ideally include soil type variation, climatic variation and various habitat types.

Within site variation

Within a site where monitoring is carried out, the variation should be as small as possible in order to reduce the number of explaining variables to take into account. This is important both when selecting the monitoring habitat and when designing the sampling within the site.

Temporal variation

Inevitably, year-to-year fluctuations in major environmental conditions such as rain, summer drought and winter frost will influence survival and reproduction of monitored organisms. Fluctuations in predator-prey cycles will also induce variation for affected organism groups. If temporal variation can be explained as, e.g., model variables, uncertainty in variable determination can be reduced. Otherwise, this variation will add uncertainty to effect detection.

Methodological uncertainty

The different methods that may be applied for monitoring purposes each have forces and weaknesses with respect to objectivity, accuracy, applicability for statistical purposes, the factors that they determine and the level at which they are able to detect change. Optimisation in choice of methods will depend on the particular questions being asked for the specific GMHP case.

Taxonomic problems

Problems with respect to taxonomy include the correct determination of species, separation of subspecies and identification of hybrids. Identification of taxa in most taxonomic groups will require highly skilled expertise, which may be difficult to obtain for some organism groups. It must be stressed that trained personnel with taxonomic skill are essential for most ecological research and monitoring of effects.

Statistical considerations   Obviously, it may be impossible to monitor extensive areas or a region (e.g., the EU). Therefore some shortcuts, which reduce the effort to an acceptable extent without disabling the established monitoring program, have to be found. Such shortcuts are created by help of statistical considerations and knowledge of statistical parameters describing the variation related to the variables that should be monitored, see Chapter 8.

Personnel and equipment   The personnel and the equipment involved in monitoring may vary over time and space. In order to reduce this variation it is important that parallel monitoring, inter-comparisons and inter-calibration programs are carried out.

3.3.2 GMHP monitoring experience

Hybridisation and dispersal   Specific experience concerning environmental monitoring of dispersal or effects of commercially grown GM plants is very limited. Several studies have demonstrated that hybridisation with close relatives do occur frequently (Jørgensen & Andersen 1994, Bartsch & Pohl-Orf 1996, Jørgensen et al. 1996, Darmency et al. 1998). Dispersal of the GM crops is likely to occur if the non-modified receiver species is already invasive, and with higher rate if the inserted trait offers resistance towards environmental stress (Steward et al. 1997, Scott & Wilkinson 1998).

General information on GM crops   For some major crops which have been genetically modified, documents covering broad aspects of biology, ecology, crop use, etc. have been made by state or international organisations: beet (*Beta vulgaris*) Højland & Pedersen (1994b), Gerdemann-Knörck & Tegeder (1997); carrot (*Daucus carota*) Højland & Pedersen (1994a); potato (*Solanum tuberosum*) Bijman (1994), Højland & Pedersen (1994), OECD (1997b); oil-seed rape (*Brassica napus*) Højland & Poulsen (1994), Gerdemann-Knörck & Tegeder (1997), OECD (1997a). This information is useful in risk assessment and for specific issues in monitoring.

Few data exist on effects of GMHP   Effects on the environment of GMHP other than regarding dispersal and the hybridisation have only been indicated to a limited extent so far (Steward et al. 1997, Hilbeck et al. 1998, 2000). However, possible environmental effects of Bt-toxin in pollen of genetically modified corn (*Zea mays*) on the monarch larvae, *Danaus plexippus,* have been reported (Losey 1999). A major concern for possible invasion of GMHP and harm to natural habitats is based on the experience of introduced species, which in some cases have invaded and caused serious impacts to the environment. In Europe, *Heracleum mantegazzianum* and *Rosa rugosa* are well-known examples of anthropogenic introductions that have displaced the native flora. Changes in agricultural practice accompanying the use of GMHP are likely to produce changes within the agricultural ecosystem. For instance may the weed populations and size and composition of the seed bank change because of changes in the use of herbicides, this may again be followed by changes further up in the food chain (see also Chapter 5).

3.3.3 Monitoring methods for other environmental purposes

For other kinds of effects to the environment, monitoring programs have been developed (From & Söderman 1997, Lawesson 2000) and some have also been implemented (UN-ECE 1998) as well as different methods have been suggested for different purposes (Jensen & Christensen 1994).

Existent information and scientific considerations decide which variables to monitor. Methodological and statistical considerations determine by which methods the selected variables should be sampled and analysed. In this process there is a number of uncertainty factors that influence the variable which should be taken into account, see below and Section 3.3.1 on methodological uncertainty.

Comparison of methods
for cover estimate

Different methods for vegetation analysis have been compared for monitoring purposes (Bråkenhielm & Qinghong 1995). They compared three methods for cover estimate with respect to sensitivity, accuracy and time consumption. Their results indicate that the method that uses visual estimate was the most accurate, the most sensitive and the least time consuming. The point frequency method was the most time consuming by approximately 40% compared to the visual estimate. Without a registration of rare species the point frequency method failed to register a large part of the species, especially in very diverse plant communities. Therefore application of the point frequency method will be further time consuming. On the other hand the point frequency method ensures that the whole area is systematically investigated and also that species growing in the lower storeys of the plant community are detected and quantified.

Cover estimates are not
consistent

In the cover estimate method, a substantial amount of subjectivity can not be avoided, especially when different observers do the estimates. This is confirmed in a study comparing the ability of different observers to visually estimate absolute and relative cover of eight Alaskan forest floor species, here it was found that observers could not do this in a consistently and repeatable way (van Hees & Mead 2000). If this method is used, it is essential that the observers are calibrated by frequent inter-comparisons (From & Söderman 1997).

Leaf Area Index as a
vegetation measure

The leaf area, LA, or leaf area index, LAI, can be used as alternative measures of vegetation density per unit of area (Kjellsson & Simonsen 1994). LA is a good indicator of individual plant vigour and competitive ability. Thereby is LA especially useful when monitoring changes in total above ground biomass of monocultures or total vegetation.

Quick and accurate in situ methods by using indirect optical measurements are available (Welles 1990, Chen & Cihlar 1995). A linear relationship between aboveground biomass and plant cover, determined by image-analysis, has been reported for species in open vegetation (Röttgermann et al. 2000).

Frequency in subplots

From and Söderman (1997) and Lawesson (2000) recommends the frequency in subplot method measured in permanent sampling sites. Concerning replicate numbers they recommend more than 20 replicates in homogenous vegetation and more than 50 in gradient studies. However, this method was found to deviate the most from the true value in its estimate of cover and it could not be correlated to the values predicted by the methods estimating closest to the true values as measured by photography and digital methods (Bråkenhielm & Qinghong 1995).

Point frequency

Other investigators recommend the point frequency method both in relation to GMHP registration (Jensen & Christensen 1994) and generally (Jonasson 1983, Jonasson 1988), because of the objectivity of this method and its applicability for statistical treatment. By counting the number of intercepts between the different species present in the vegetation and a thin pin in a grid, measures of species frequency and cover are provided. The possibility of non-destructive biomass estimates (Jonasson 1983, 1988) is also in favour of the point frequency method, because this may be an important parameter for the monitoring of effects of the GMHP on plant community level. The consumption of time speaks against the point frequency method.

3.4 Factors that may make a community invasible

We have limited knowledge on which types of ecosystems are the most susceptible to invasion (i.e., invasibility), and it is still very unclear how this should be assessed. Neither do we know at which level of invasion pressure, the recipient ecosystems are severely affected.

Invasibility of grasslands

Some evidence exists indicating that grasslands have a high invasibility (Mack 1989, Tilman & Downing 1994, Tilman 1997), see also Chapter 4. As some types of Danish grasslands are very species rich and sustain numerous non-indigenous species it is likely that this type also in Denmark have a relatively high invasibility, see Chapter 4.

Important factors for the invasibility of ecosystems in a multidimensional system

Primary and secondary factors believed to be of importance for the invasibility of ecosystems include time since introduction, stress, nutrient status, plant cover, leaf area, and biomass (Figure 3.2). These variables are termed secondary in this context, because they are determined by a combination of the primary factors nutrient status, disturbance and time. By determination of the most important primary pararameters for the invasibility it is theoretically possible to obtain a measure of the invasibility in a multidimensional ecosystem. The box in Figure 3.2 symbolise a plant community at an arbitrary place in the multidimensional space formed by time and environmental conditions. The arrows indicate that the community can move in different directions dependent upon the direction of the succession and environmental forces acting on the system.

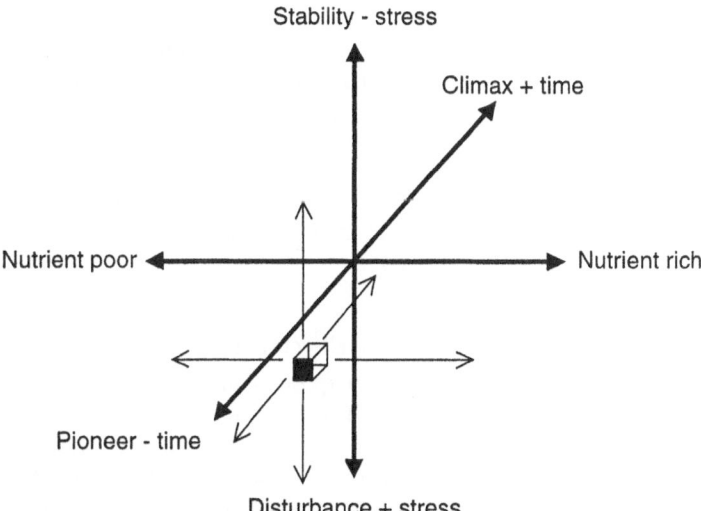

Figure 3.2 Primary properties of an ecosystem or a successional system of ecosystems of importance for its invasibility. The box represents an average with a variation in three directions in a three dimensional space of environmental conditions.

Need of analysis method   Some of the variables may be problematic because they consist of more independent variables such as, e.g., the nutrient status. In order to attain a measure for invasibility it may be necessary to construct a system that allows the different variables to play together.

3.5 Suggestions for a tripartite monitoring program

In the following detailed suggestions for the monitoring program consisting of three subprograms are presented.

3.5.1 Dispersal subprogram

Transgene dispersal   Monitoring for presence of the transgene insert can be carried out with very specific DNA methods developed especially for the purpose of detection of the specific transgene (see Section 6.2). This part of the monitoring should be based on the species represented by the GM crop or hybridisation-partners of relevance. The properties of the transgene and the host crop species decide which species are of relevance in the monitoring program. The monitoring takes place by following selected populations of the same plant species or potential hybridisation partners. The frequency of the transgene could either be investigated by DNA-sequential techniques, by marker genes or possible by screening for the transgene (see methods reviewed in Kjellsson et al. 1997), e.g., herbicide resistant plants can be identified by use of dot tests for herbicides, frost resistance by freeze trials, etc.

Important factors for the dispersal of the transgene include: fitness of the transgene crop, occurrence of compatible wild relatives, proximity of compatible wild relatives, fitness increase by the transgene insert and probability of hybridisation with wild relatives.

Weediness

The transgene could confer weediness to the host and therefore within field monitoring should be included as part of the transgene dispersal monitoring (see Chapter 5).

3.5.2 Effects subprogram

Large-scale investigations in permanent, extensive areas

The purpose of the effects subprogram is to detect effects of the GMHP or its use on the environment in form of altered ecosystem function or composition. It is suggested that this part of the monitoring program is established on permanent long-term, large-scale investigation areas. These areas are to be monitored continuously in plots laid out at different distances from fields where GMHP crops are grown. One problem, which has to be dealt with, concerns the practice of crop rotation in agriculture. However, this may be solved if monitoring areas are placed within regions where specific crops are grown in large quantities.

We suggest that areas for monitoring should be placed on governmental properties or on areas provided by the GMHP industry. These monitoring areas will be required for a period of minimum 10 to 20 years. Such large-scale investigation areas should be established in different parts of the EU when the ecological risk assessment points out a need for effects monitoring and the GM crops have been approved for commercial use. Probably at least ten places in the EU will be the minimum number needed in each particular case.

Causal relationships for monitored effects difficult to attain

The effects monitoring subprogram is most likely to detect some effects to monitored organisms or environmental parameters. However, any causal relation between observed effects and the invasion of a GMHP will be difficult to test and in most cases even to render probable. A baseline of data from a period before monitoring started, makes effects detection much easier. The use of controls during a period of monitoring can partly substitute a baseline if similar areas are available (Table 3.1).

Table 3.1 Overview of probability of effects detection with different monitoring alternatives. A: Effects monitoring in large-scale investigations; B: Monitoring for effects when hybridisation partners have been detected, baseline determined in plots where hybridisation partners occur; C: Effects monitoring in plots where GMHP dispersal or hybrids have been detected, no baseline detection in this approach.

	A Large-scale + baseline + controls	B Variable scale + baseline - controls	C Variable scale - baseline + controls
Effects of use of GMHP	High	Medium	Medium-low
Early warning	(+)	(+)	(+)
Effects on soil microfauna	+	(+)	(+)
Effects on soil micro-biology	+	(+)	(+)
Effects on insects	+	(+)	(+)
Effects on birds and mammals [a]	(+)	-	-
Effects on soil processes	+	+	-
Time and man power demand	High	Medium-High	Medium
Expenses	High	Medium-High	Medium

[a]: Effects are possible if large-scale use is applied and particular birds and mammals forage and depend on organisms, such as insects, that are negatively affected by the GMHP. Toxic compounds in target animals, e.g. insects, may possibly accumulate in higher predators.

A time and cost reducing procedure could be, to start monitoring for dispersal and then look for effects only if significant dispersal is detected. In this way the effect monitoring will be restricted to cases where there is a good chance to observe and explain effects. The drawback of this approach is that the chance of observing effects in more organism interactions (e.g., food webs) is diminished. Furthermore, effects not directly linked to the spread of the GMHP are not likely to be identified. However, direct effects should be monitored first, as these are more likely to happen and to be adverse.

Recommendations for decisions on monitoring during risk assessment

We recommend that during the ecological risk assessment, decisions are taken based on ecological consideration whether effect monitoring should be initiated, and that suggestions are given for relevant methods and procedures in each case. As minimum requirements, Biotech farming should always require monitoring of effects directly in the agro-ecosystem where the GM crops are grown and in the close surroundings.

3.5.3 Surveillance subprogram

The objectives of this part of the monitoring program are meant to detect unforeseen effects. However, unforeseen and unlikely effects can also be detected in the 2nd. part of the monitoring program, which is designed to detect environmental effects. Therefore, it is suggested that the main purpose of the surveillance subprogram is to give an overview on a wider scale. Ideally, surveillance should take place everywhere where GMHP are grown or transported. The idea is that everybody who handles viable seeds of GMHP receives instructions of what to look for and where and how to report observations of possible hazards. Observations should include: occurrence and increases of GMHP volunteers, and if possible their near relatives, in cultivated fields and surroundings (see Chapter 5). Occurrences in unexpected places, such as wasteland, harbour surroundings, feeding stuff trades, roadsides, etc., should also be surveyed. In this way, an overview of unforeseen dispersal caused by the transport and cultivation of the GMHP could be obtained on a wider scale.

3.5.4 Problems related to effect monitoring

Lack of knowledge and lack of accept criteria

As indicated above, the aim of monitoring for environmental effects from GMHP raise some methodological problems that have to be dealt with. Other problems concern the lack of knowledge on, e.g., organism interactions and the lack of relevant accept criteria which are also needed in risk assessment and risk management. However, the field of risk assessment of GMHP is relatively young and accept criteria still have to be agreed upon (Kjellsson 1997, Strandberg et al. 1998, Kjær et al. 1999).

3.6 Data analysis and evaluation of results

Data types from monitoring

Different types of data will be available from monitoring of effects which are presumably caused by the GMHP. For changes in the composition of the ecosystem, these include cover and biomass of plant species. For other parts of the ecosystem, such as invertebrate populations, they include population size and diversity (see Chapter 7).

Multivariate data analysis

The methods for data analysis presented here include multivariate statistical methods (ter Braak 1996, Lawesson 2000) see Figure 3.3, analysis of variance and analysis for correlation. When multivariate statistical methods are used, the data are arranged along multiple ordination axes, which represent changes in the data set, such as vegetation gradients.

Ordination score for each plot

Each plot is assigned a score on each axis, which represents its placement along a number of gradients/ordination axes. Lawesson (2000) presents different ordination methods, such as eigenanalysis-based ordination methods and distance based ordination methods. The environmental variables to be analysed by use of analysis for correlation include distance from field, frequency of the GM crop and level of disturbance. Differences between control plots and influenced plots and between sites that relates to measures such as LAI, cover, biomass and biodiversity are suggested analysed by use of ANOVA.

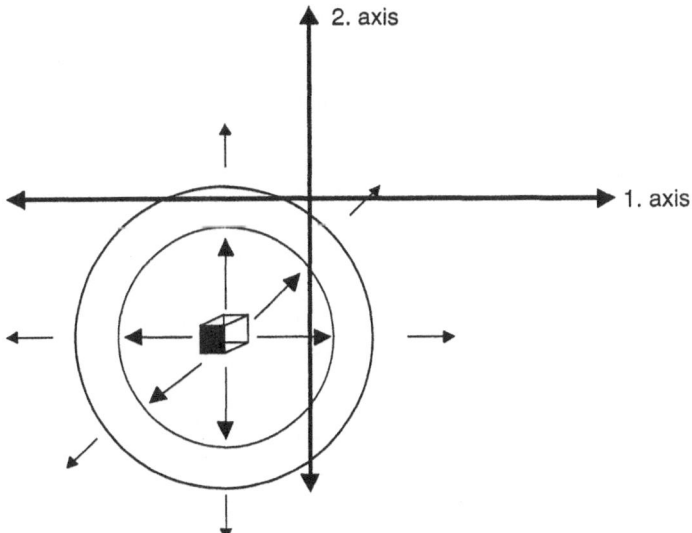

Figure 3.3 Schematic representation of how to use multivariate statistics to determine plant community changes. By use of multivariate analysis, such as "Detrended Correspondence Analysis" (DCA), a sample of species data from a monitoring location can be characterised (located) in a multi-dimensional space (the box). The inner circle describes the limit for what could be called "not alarming variation", whereas the outer circle describes the limit for changes beyond what would be expected in unaffected nature. The space between the two circles should be interpreted as an interval of insecurity. Consequently, species communities which are found in this space should receive special attention.

Cause-effect relationship and the use of control-plots

When changes to a plant community are observed it may, however, be problematic to establish or even provide circumstantial proof of a cause-effect relationship between a GMHP and the community changes. It is important to determine a set of environmental variables that could explain different kinds of changes. Unaffected control-plots could serve as an important tool to assist in detection of specific cause-related effects. Hence, if only the exposed plots are affected, the likelihood that the effects are due to the GMHP is higher. Ideally control-plots should be established in such a way that the only factor separating these plots from the GMHP-monitoring plots is, that they are not affected by dispersal or use of the GMHP. This, of course, is only possible to a certain degree, because the controls may show great variation.

Permanent plots at distance intervals

However, an approach suitable for the purpose can be established by use of permanent monitoring areas, where permanent plots are located at different distance intervals from the field where the GMHP are grown (Figure 3.4). In order to ensure adequate intervals of time for the monitoring process, it is suggested that the permanent plots are established on non-cultivated large-scale investigation areas. In this way it is avoided to confer restrictions on the growers of the GMHP other than those set by the ecological risk assessment.

Graphical illustration of monitoring outcome

The primary outcome of monitoring according to the scheduled method gives a result represented by the box in the centre of the circle (sphere in a three dimensional room) in Figure 3.3. The placement of the box in the centre represents the baseline of the ecosystem, which is the result from the initial monitoring. The arrows and circles represent natural variation and/or changes detected as a result of the monitoring carried out later.

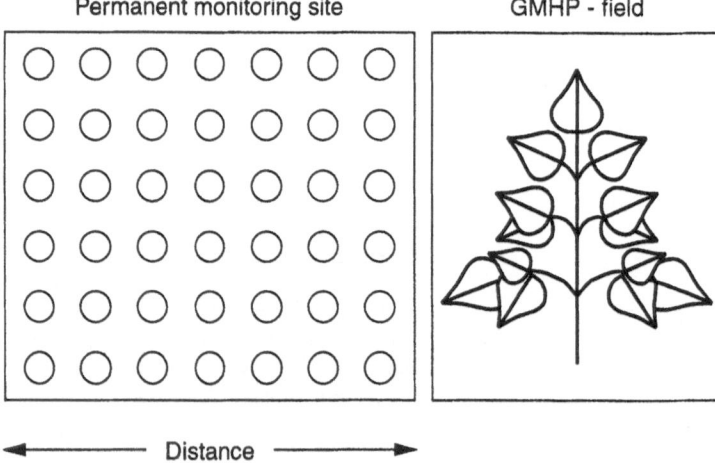

Permanent monitoring site   GMHP - field

◀――――――― Distance ―――――――▶

Figure 3.4 Suggested arrangement of a permanent GMHP monitoring site situated in large-scale investigation areas. Sites will be different kinds of exposed sites located close to fields with GMHP cultures. Within each site some 40 permanent plots are situated and it is within these the baseline determination and the monitoring for effects takes place.

Significance of change

When permanent plots are monitored over a long period of time, some changes of the ecosystem are likely to take place. The ordination score is a measure giving a profile of each plot and as such, it is well suited as an object for test of significance of change. One or k sample statistical test procedures can be applied to test the significance of the change depending on whether single or multiple species/measures are included.

3.7 Extent of monitoring

Specific design in each case of post market monitoring

Each application and approval for placing on the market should give rise to a specifically designed monitoring program. Both the properties of the inserted gene and the properties of the receiver plant should be taken into account, when it is decided where and to which extent monitoring should take place. In first hand the design is mainly a result of the risk assessment. Later, the design may be altered if the monitoring program points out unforeseen trends or shows up to be inadequate.

Surveillance of transport dispersal

GMHP should always be surveyed where transport of genetically engineered seeds takes place, in harbours, along main roads and at places where the products are reloaded etc., see section 3.5.3.

Transgene dispersal, effect and surveillance

The dispersal- and effect subprograms are suggested to be carried out in sites located near fields where GMHP crops are grown, for habitat selection, see Chapter 4. Some aspects of within field monitoring for dispersal of the transgene insert could be handled by the surveillance sub-program others could be treated in field monitoring (see Chapter 5). Monitoring for effects of the GMHP crop within the field, for instance effects on the weedy flora, soil and soil organisms (see Chapter 5 and 7) should be designed as parts of the effects monitoring sub-program.

The risk assessment forms the premises for the monitoring design

The design and characteristics of each monitoring program is mainly the responsibility of the assessment authority, because monitoring should be carried out in accordance with the risks identified during the assessment of the GMHP. The ecological assessment forms the basis on which the relevant nature types for monitoring should be selected. The surveillance part of the monitoring program should be carried out wherever viable seeds of GMHP are handled, from distribution to field use and crop rotation (see Chapter 5). For each post marketing case, a particular monitoring program should be designed and initiated.

Period of monitoring

Ideally, the monitoring should be carried out in association with the same GMHP for a number of years and crop rotation cycles, minimum five to 10 years and with 2 to 3 rotation cycles. This may be difficult to accomplish, e.g., because of restrictions made to the land use by the farmers and the costs involved. Even so, it is essential that continuity of data collection is maintained. Otherwise, it may be difficult to reach any conclusions on the potential environmental effects of the GMHP. In the new amending EU-directive, 90/220/EEC, a reassessment of consent for marketing must be done after a maximum of 10 years, which should be sufficient for initial monitoring and detection of possible adverse effects to the environment.

Different subprograms for different types of GMHP

The three different parts of suggested monitoring program can be implemented alone or together depending upon the properties of the GMHP and the conclusions of the risk assessment. A subprogram that provides information which is nice to know but not essential for the particular types of hazards expected, could be cut. For a herbicide-resistant crop for instance, monitoring of effects outside field areas and surroundings can often be omitted is hybridisation do not occur (see Table 5.1). In other cases, such as biotechnological induced changes in the colour of a flower, this may be believed to be harmless to the environment, but could affect the wild pollinator fauna, if the transgene is dispersed. A sexually reproductive GMHP with altered growth properties or increased stress tolerance would require a comprehensive monitoring and use of all three subprograms.

Extent of monitoring based on precautions and responsibility

In cases where the possible effects are less obvious, the considerations are less easy to take. In the end, only practical experience from an extended period of observations can tell us what to expect. The safe choice is to use the precautionary principle and choose the most comprehensive monitoring program. However, the cost and labour involved, will require that resources are primarily allocated to monitor effects in areas where exposure to the GMHP or the transgene is high. The final decisions of content and extension of monitoring programs are always the responsibility of the regulatory bodies of the ecological risk assessment.

3.8 Reassessment of established monitoring procedures

No adverse effects detected

After the GMHP monitoring program has been established and a period of time has elapsed, different kind of results may be experienced. Hopefully, no adverse effects are detected. If this is the case after a longer period, say ten years, it is likely but not proved that the risk assessment of the GMHP that led to the commercialisation and introduction was correct. In this case, adjustment of monitoring procedures can be made, resulting in, e.g., larger time intervals between monitoring or that only a general surveillance program is maintained and the approval for placing on the market is continued.

Unacceptable effects detected

If unacceptable effects are detected, whatever they may be, the situation is different. The monitoring program has detected significant adverse effects to the environment, which the risk assessment has not been able to assess correctly. Hence the assessment has to be reevaluated and probably scientific trials have to be initiated to further identify the causal relations between GMHP and effects. Also the consequences of keeping the GMHP on the market has to be clarified. As the effects are termed unacceptable, the most likely step to take is to stop the placement on the market and the agricultural use. From the revision of the risk assessment it also follows that GMHP with properties similar to the one that was stopped should be reassessed. The risk assessment procedures may also need to be evaluated and perhaps revised.

Effects detected from surveillance

A third situation occurs if the surveillance program detects effects, which have not been detected in the monitoring program. In this case the monitoring program as well as the risk assessment should be adjusted. Furthermore, the results from the surveillance program should be analysed with reference to need of further experimental data and to the fate of the GMHP on the market.

4 Perspectives of habitat selection

In Chapter 4 questions concerning the choice of monitoring habitats are discussed. It is argued that the choice in principle shall be the result of a "case by case" assessment, which in fact is what should be processed during the ecological risk assessment.

Invasiveness factors

Factors are discussed, that determine the invasiveness of different land use types. The discussion is carried out with Denmark as an example, but the principles used should apply in a wider context too.

Disturbed areas are suitable targets for early warning

Disturbed areas with low vegetation and high abundance of herbs and grasses are identified as suitable for monitoring purposes. Firstly because they are widely distributed and often are found in connection with more intensively cultivated agricultural areas. Secondly these vegetation types are interesting because they include roadsides, ditches and edges of fields, which are often the first areas to receive stray GM seeds. In this way, disturbed areas represent an important kind of biotopes, which can provide an early warning for adverse effects.

Invasion in natural grasslands

The types of natural grasslands that are most likely to be exposed to invasion of GMHP crops are probably those with low species richness on poor dry sandy soils and those with high species richness on dry calcareous soils. Both types are among the most original nature types in Denmark. Because of their dryness and special soil types tree growth and forest development here is slow or not occurring.

Comparisons with information from other monitoring programs

Furthermore, it is debatable whether any effects other than dispersal of the GMHP or the transgene can be ascribed directly to the GMHP or its use. The difficulties are that a lot of different environmental impacts affect the ecosystem independent of the GMHP. Comparisons with control plots and information from other environmental monitoring programs may possible help to identify the origin of monitored effects, but it is doubtful whether this will be enough. Cases where adverse effects caused by the GMHP are suspected should be followed by explicit hypotheses of causal relationships and tests should be made to accept or reject these hypotheses.

4.1 Background for habitat selection in Denmark and EU

The Danish land area is a heterogeneous mosaic of more or less anthropogenically influenced areas. Cultivated land, grass-dominated areas within rotation etc. covers the mainpart approx. 62%. Ranked in decreasing order the other land use types cover; forest approx. 12%, urban areas, roads etc. approx., 12 %, meadows, marshland etc. 6%, heaths, dunes and bogs 5%, hedgerows, ditches etc. 3%, lakes and streams 1% (Danmarks Statistik 1997). Except for some of the lakes, raised bogs and coastal areas, no area in Denmark is totally free of direct influence from human activities. These influences range from effects from water uptake and farming to deposition of nitrogen and other substances from air-pollution, which may lead to plant community changes.

Monitoring and habitat selection may vary in EU

The EU countries cover a broad range of habitat types ranging from coastal habitats to boreal forests. A description of the different types of natural habitats existing in Europe is available from the EU (Romao 1996). The description includes priority and non-priority habitats, geographical distribution and lists of main plant species. The crop species used also differs from the mediterranean climate zone to the northern temperate zone. Thus, crop species having no near relatives in Denmark may have relatives in other parts of Europe where cultivation also takes place. This will have consequences not only to the habitat selection, but also to the monitoring design, which may vary according to the regional differences.

Areas exposed to invasion of crop species

Theoretically, invasion by crop species can take place in most of the terrestrial area, except the main part of urban areas etc. However, invasions of agricultural crop species are also quite unlikely in forests, but obviously, GM trees and bushes etc. will be able to invade. Open areas are to some degree exposed to invasion by GM agricultural crops and to the transgene through hybridisation with related species. Which ecosystem a specific crop species is most likely to invade depends upon the ecology of the crop species. Hence in order to decide where to monitor for dispersal it is necessary to know the ecology/life strategy of the crop species. Or as cited by Levin (1990): "For any individual introduction it is the characteristics of the engineered organism and its recipient environment that should be the base for the risk assessment".

Habitat description and invader ecology

The subject of habitat selection for monitoring purposes was superficially treated in Chapter 3; the present chapter provides more details. A firm measure of habitat invasibility would be a valuable habitat describing variable to know in the selection of the habitats in which to perform the monitoring. However, as already stated very little is certain in relation to determination of invasibility. Furthermore it should be noticed that habitat invasibility always depend on the particular adaptive traits of the invasive species, such as life-history, phenology, stress-tolerance, etc.

4.2 Important factors for selection of exposed areas

Which factors determine plant invasion in exposed habitats? Answers to this question have not yet been fully determined. Several factors may be important prerequisites for an invasion to take place, such as disturbance level, vegetation cover, plant diversity and the nutrient status of the ecosystem.

Disturbance and invasion of ecosystems

Different kinds of soil disturbances are most frequently mentioned as a prerequisite for invasions to take place (e.g., Crawley 1987, Hobbs 1989, Rejmanek 1989, Crawley & Brown 1995). The major types of natural disturbances include fire, grazing, wind-exposure of soil, tree-fall, etc. Man-induced disturbances, such as tree cutting, cultivation and farming, have more or less severely affected many ecosystems in historic time. Disturbance creates the space necessary for establishment of the invader. Many agricultural crop species (e.g., oilseed rape and cereals) have an annual and biennial life-cycle, which naturally constitute the early successional phases of disturbed ecosystems. Rejmanek (1989) states that basically, the probability of successful invasions seems to crucially depend on the extent and type of disturbance, on the number of non-native species propagules deposited in the community per year, and how long the community is exposed to import of propagules.

Biomass and vegetation cover determine community invasibility

The amount of biomass or cover may be the most efficient indices of community resistance in some situations, i.e., the higher the cover and biomass of the recipient community the higher the resistance to invasions.

Disturbance and soil turn-over

Often disturbance in itself is not adequate to ensure a good establishment. Results seem to show that when the disturbance is of a kind that also buries the seeds in the soil, germination and establishment are improved. Most seeds require burial in the soil for protection against predators and for optimal germination conditions. In nature such burial may be performed by trampling from ruminant grazers or by animals such as voles and earthworms. The combination of molehills and trampling may create a very good basis for establishment of stray seeds from agriculture.

Distance and the probability of pollen and seed dispersal

Distance is another important factor, because the density of pollen and seed usually decreases with increasing distance from the dispersal source (e.g., the initial invasive GMHP population). The probability for dispersal depends on the particular dispersal vector, such as wind, insects, animals and man. The probability of successful invasion further depends on the probability of seed germination, survival and reproduction (i.e., recruitment of new individuals) and population establishment or on the formation of fertile hybrids. The distance of the GM crop to wild populations and to relatives of the crop species becomes important for estimating the risk of hybrid formation. In conventional farming of certified seed crops, different isolation distances between fields are required for each crop species (normally from 200 to 1000 m's depending on species). However, these distances do not totally exclude pollen dispersal and hybridisation but sets low realistic limits for the formation of hybrid seeds. This has recently become a problem to organic farming because total seed purity from GM material cannot be guarantied if GM plants of the same species are cultivated in the area or even far away (Moyes & Dale 1999, see also Chapter 5). It is worthwhile to note that not only the conditions in or near the field may be important. Crawley and Brown (1995) found that the transport of rape seeds to the crushing plant was more important for the occurrence of oil-seed-rape along the road than the presence of cultivated oil-seed rape fields nearby.

High diversity of native plants does not exclude invasion

A survey of global plant invasion studies showed that communities, which were rich in native species had more, not fewer, exotics (Lonsdale 1999). This is contrary to the most widespread theory of how native floristic species richness influences site invasibility (e.g., Tilman 1997). However, Stohlgren et al. (1999) also found that meadows and forests with a high native floristic diversity on rich soils had a high exotic species richness. In grassland habitats they found that the sites with the lowest native floristic diversity had the highest abundance of exotic species, which is in accordance with the above-mentioned widespread theory. Ejrnæs (pers. comm.) expresses the opinion that in Danish natural grasslands, the highest plant invasibility is found on dry sandy soils with low species richness and on dry calcareous soils with high species richness). This observation supports the view that simple rules, which are generally valid for ecosystems in many different regions, are difficult to establish.

Soil nutrient status may influence on invasion

The soil nutrient status is also important for the establishment of new plants, although in a complex way, because high nutrient levels may be a prerequisite to nutrient demanding species, such as is often the case with crop species. However, GM crops may be altered to be competitive in less favourable environments with respect to nutrients or their genes may be transferred to relatives with other competitive properties and thereby further increase their fitness. In this context it is interesting that a study showed that heaths and other dry habitats were most prone to invasion of all habitat types in Southwest Denmark (Andersen 1997).

Phases of invasion

Invasion of exotic species can be split into three phases: introduction, colonisation and naturalisation (Heywood 1989). Only when these three phases are accomplished, the invasion can be termed successful. Native species can become invasive in new habitats if disturbances or changes in management lead to altered environmental conditions (Thompson et al. 1995). Often the process of invasion is delayed. The invasion can stop or be delayed in the initial establishment phase because it takes time for the right combination of properties of the invader and the environment to occur. Furthermore, the established invader has to accommodate genetically to the environment or wait for the environment to change in a more favourable direction. This change can take place from man-induced management, pollution or as a result of natural processes.

Frequency of disturbance is an important factor for invasion

The most important property of the receiving ecosystem in relation to invasion seem to be disturbance, meaning that the space necessary for establishment is offered, and that the recipient ecosystem is situated such that the event of dispersal occurs frequently.

4.3 Background for habitat selection in relation to GMHP monitoring

In selecting habitats for monitoring purposes, knowledge on certain characteristics of the GMHP and the ecosystems may help to decide where to monitor and for what. In each case an ecological assessment of the GMHP include information on its ability to survive, establish, naturalise, spread, invade and hybridise in the region where it has been approved for agricultural use. Hereby follows that the GMHP may show different aspects within a larger region, such as the EU. The ecological conditions vary and an ecological risk assessment valid for one area may be invalid for another area with another set of more or less extreme environmental conditions. Therefore the ecological risk assessment should pay attention to the whole span of environmental conditions within the region where the assessment is valid.

Monitoring decisions

In the process of deciding on monitoring design, a checklist of the autecological characteristics of the GMHP and the unmodified parental species may prove useful. This should also include regional ecological characteristics such as temperature regime and presence of suitable ecotypes, see Table 4.1.

By creating and filling out such a table, profiles of the GMHP and ecosystems of relevance are identified, which may help in deciding where to monitor. This will not be an easy task - and will require detailed information on plant ecology which is partly available in previous reports published by the Danish National Forest and Nature Agency and OECD (e.g., Duhn 1994, Højland & Pedersen 1994, Højland & Poulsen 1994, OECD 1997a, 1997b, 1999a, 1999b). The identification of data need in a tiered approach to ecological risk assessment has been thoroughly described (Kjær et al. 1999). Therefore, an identification of particular risks of a GMHP and exposed areas for monitoring can ideally be deduced from the ecological risk assessment.

Table 4.1 Environmental parameters in relation to GM plant characteristics. A preliminary checklist for guidance in choice of monitoring habitats for GMHP. Plant characteristics include both natural and biotechnological induced traits.

Process	GMHP characteristics	Site and environmental requirements
Sexual reproduction	Type and strategy: pollination, seed dispersal and seed bank	Reproductive potential available (e.g., pollinators, dispersal agents, etc.)
Hybridisation	Genus, formation of known hybrids (see Table 6.3)	Occurrence of relatives in different types of sites
Vegetative reproduction	Tillers, tubers, etc.	Environmental conditions available for formation of vegetative structures
Plant-environment interactions	Cold and frost sensitivity or tolerance	Temperature regime, occurrence of frost, exposure to extreme environmental conditions
	Heat sensitivity or tolerance	Temperature regime, exposure to extreme environmental conditions
	Drought resistance or sensitivity	Drought regime (e.g., summer drought), soil water balance
	Humidity sensitivity	Water regime (e.g., winter rain), soil water balance
Plant-organism interactions	Pathogen resistance or sensitivity	Occurrence of target pathogens (vira and fungi)
	Pest resistance or sensitivity	Occurrence of target pests and nontarget herbivores (e.g., insects, soil arthropods)

Ellenberg index

Indices like the Ellenberg index of ecological spectra of plant species (Ellenberg 1974, Ellenberg et al. 1992) may prove useful for the purpose of habitat or regional preferences (Thompson et al. 1993). However, the Ellenberg index should be used with some caution; for instance the Ellenberg N-values have been found to correlate only weakly with soil parameters including nitrogen – instead a strong relation to biomass production has been found (Shaffers & Sýkora 2000). Additionally, a numeric database, DANVEG, for Danish grassland types can be used (Nygaard et al. 1999a). This software uses a sample of species as input for decision upon the most likely parent grassland plant community.

4.3.1 Case-by-case selection

General guidelines for the selection of monitoring habitats are presented in the section above. In the actual case of approval of a new GMHP for placing on the market, a case-by-case selection of monitoring habitats, based on the species-specific ecological assessment, necessarily has to be carried out. This task is part of the ecological risk assessment (EU annex VII, Kjær et al. 1999).

4.4 Which habitats should be monitored?

In principle, no nature types can be excluded from monitoring. However, some types are naturally more exposed to invasion than others (e.g., disturbed habitats, heath). At least the surveillance part of the monitoring program always include some kind of monitoring activity in nature areas close to GMHP cultivation areas, see Section 3.5. The dispersal- and effects subprograms monitor on selected sites in cases where the ecological risk assessment points out a need.

Environmental stress, pests and pathogens

The nature types to be included in the monitoring activities depend upon the ecological characteristics of the GMHP. Inserted properties offering tolerance to different natural stress types such as frost, salt and drought will indicate the habitats which are in risk of invasion from the GMHP (e.g., seashores and marshland in case of salt-tolerant plants). Furthermore, resistance to attack from insects, fungal diseases, viruses and bacteria will help decide in which habitat type to monitor. For instance GMHP with drought resistance are most likely to establish in dry habitats. Consequently the monitoring activities are recommended to take place in dry type habitats, for instance dry grassland, maquis, steppe and heathland. The existence of possibilities for hybridisation necessarily influences the choice of monitoring habitats.

Anthropogenic stress

Similarly resistance or tolerance to herbicides indicates that the monitoring should be carried out in habitats where this kind of anthropogenic (artificial) stress is exerted, which most often will be in cultivated fields. Other habitats subject to herbicide use include roadsides, private forests, nurseries, courtyards, gravel roads and areas adjacent to these because of spray drift of herbicides from application (Marrs et al. 1993, Marrs & Frost 1997). It is possible that areas subject to influence from long range transported herbicides should be considered too, but probably the effects are too small (Klepper et al. 1998) to give herbicide resistant GMHP a significant advantage.

4.4.1 Why select disturbed and grass dominated habitats?

Disturbed areas and grass dominated areas

A potentially large variety of biotopes could be selected for monitoring GMHP dispersal and effects of biotech farming. No biotope can be omitted *a priori*, as stated in the beginning of the chapter. If the ecological risk assessment points out a certain habitat as potentially exposed, we certainly recommend that it is included in the monitoring program. Temperate agricultural and urban cites are among the most invaded biomes (Lonsdale 1999) and should be considered (see Chapter 5). However, disturbed areas and areas dominated by grasses are through their distribution both on a global and a local scale well suited as the "default" monitoring habitat at least in temperate regions.

Natural grasslands versus locally important habitat types

In many regions on the continents, temperate grasslands have been pointed out as a vulnerable type of biotope (Mack 1989, Tilman 1997, Stohlgren et al. 1999). In Australia and parts of North and South America, temperate ecosystems have experienced invasions, which completely have replaced the native flora of its grasslands (Mack 1989). This has not happened in Europe, where extensive grasslands primarily occur in the continental part of Eastern Europe. In some other regions of Europe, grass dominated areas may not be an obvious choice for monitoring. Especially in countries with mountain areas, coastal dunes, heathland and maquis etc., the locally dominant habitat types may be more important and obvious choices.

4.4.2 Danish disturbed habitats dominated by grasses

Definition of disturbed areas

Disturbed grass-dominated areas, *sensu lato*, can be defined as low vegetation habitats dominated by grasses and forbs, where shrubs can either be present or absent and trees are absent or rare. Grazing or cutting is most often necessary to avoid succession towards shrub or forest. The type includes meadows, upland areas, roadsides, ditches and similar habitats in this broad perception of the vegetation type.

Early warning in disturbed grassland

In Denmark disturbed grass-dominated areas are widespread. Moreover they are not confined to certain regions or climates. They are found on almost all soil types and are very often found in connection with intensively cultivated arable land, where GM crops are expected to be grown. These properties may prove valuable in providing test areas for early indication of potential problems from cultivation of a new GMHP.

4.5 General changes affecting plants and other organisms

As monitoring takes place over a long time period, some ecosystem changes will inevitable be observed. In each case it has to be decided whether the observed change is an effect of the GMHP or its use, a result or natural ecosystem changes, such as vegetation succession, or caused by some other environmental impacts. Therefore, comparisons with results from other monitoring programs may clear some of this confusion. However in many cases it may be necessary to develop hypotheses explaining the observed changes and then test these hypotheses.

Human influence on ecosystem processes

Agriculture, industry and traffic are activities of the human society that influences ecosystem. Impacts take place through the use of land for roads, buildings, agriculture, etc. Some of these impacts can be analysed and today they are well regulated. The changes they cause may sometimes be distinguished from other changes, such as those caused by GMHP, by their direct effects on neighbouring habitats. Impacts caused by air pollution from traffic, industry and agriculture are less easy to distinguish from effects of GMHP because they evolve slowly and the knowledge on how ecosystems are affected is in many cases limited. These effects tend to harm some species and favour others, either by eliminating more sensitive species, e.g., from herbicide application or by stimulation of a group of species, for instance nitrogen demanding species, e.g., from the use of fertilisers or from nitrogen deposition.

Nitrogen deposition affects the flora

In Southern Sweden it was found by use of Ellenberg indicator values (Ellenberg et al. 1992) that nitrogen was the single factor that correlated best to observed changes in the flora during a period of 58 years (Tyler & Olsson 1997). Other research generally support the above. Hence, plant species richness decreased along an increasing gradient of soil nitrogen in a Minnesota old field after 5 years (Wilson & Tilman 1991). However, really convincing material still needs to be produced. The problem is that controlled experiments are often of too short duration to produce any effects, and that observations in nature only provide indications rather than scientific proofs. Of course, direct application of commercial fertilisers in nature areas immediately produces obvious floristic changes. However, when it comes to the effects of nitrogen deposition, the changes are more slow to develop and difficult to separate from other environmental variables.

Effects of N-deposition in grasslands

Ejrnæs (1998) studied Danish grasslands in a comparative study and did not reach decisive conclusions regarding effects of deposition of nitrogen. Wilson et al. (1998) increased N-deposition experimentally on British calcareous grassland but found only limited effects as long as the grasslands were also P-limited.

Effects of N-deposition in forests

Tybirk and Strandberg (1999) studied the floristic changes in oak forest from 1916 to 1995 and found that they correlated to an increased deposition of nitrogen. However, light availability decreased in the same period as the forest grew older and the canopy became more dense. This was probably the main explanation for the observed floristic changes. This illustrates that even simple causal relationships are difficult to establish and that the complexity of nature often will require the use of elaborate statistics or modelling.

Effects of N-deposition in raised bogs and aquatic environments

From a combination of experiments and field studies Risager (1998) reached the conclusion that nitrogen deposition exceeded the critical load of the raised bogs in Denmark and already had affected the flora of the bogs. This had happened by impeding the ombrotrophic *Sphagnum* species and providing nutrient for invasive plants alien to the raised bog community. The study may be the most convincing evidence for effects of nitrogen deposition on the flora in the terrestrial environment. In the aquatic environment, increased runoff of nitrogen and phosphorous from surrounding farmland has since long been known to cause dramatic changes. Both the composition of the algae flora and submersed plants, especially those forming rosettes on the bottom, such as *Isoëtes spp.*, are affected.

Herbicide deposition may affect some species in the flora but nitrogen deposition is more important

Application of herbicides in nature areas is rare. However, many herbicides are subject to atmospheric transportation and thereby small amounts are deposited in nature areas. In lack of direct measurements of possible effects, a model-study has shown that Dutch nature areas receive 0.02 dose equivalents per year (Klepper et al. 1998). This deposition was calculated by use of a dose-response model, which also indicated that 2% of the species were affected over their NOEC (i.e., No Observed Effect Concentration). The Danish consumption of herbicides is similar to that of the Dutch (Miljøstyrelsen 1998). However, effects of herbicide drift are small outside cultivated fields compared to the effects from fertilisers and other types of nitrogen deposition (Bichel Committee 1999).

Other pollutants

Tropospheric ozone (Johnsen et al. 1991, Stewart et al. 1996), heavy metals and xenobiotic organic compounds are other types of pollutants that may affect the floristic diversity of ecosystems. In this way they may influence monitoring for other purposes, such as effects of GMHP.

Global change

Global change, caused by atmospheric pollutants such as ozone, may also affect results from monitoring in a way that can be difficult to separate from GMHP-related effects (Stewart & Potvin 1996). This is especially the case when long-term effects are considered.

4.6 Baseline definition

The basic state of an ecosystem is determined by a number of different primary parameters, some of which are shown in Figure 4.1. The characteristics of the species determine where in the multidimensional space of different environmental conditions its chances of invasiveness is the largest.

Basic state of ecosystem with succession and variation

Many North European ecosystems change over time because they are part of a succession. Unless halted, the vegetation will with few exceptions, change towards forest, which is the most widespread major type of ecosystem in Europe. In order to determine the basic state we have to be aware of species turnover, the inevitable successional changes and the built in seasonal variation due to climate and the dynamic nature of ecosystems with living species competing for resources. Some of these variables are difficult, others impossible to determine. Therefore, the basic state of the ecosystem will have to be determined by the structure, composition and variation of its plant communities. This may be done by vegetation analysis with an adequate number of replicate determinations of cover and frequency of individual species as well as total cover and biomass.

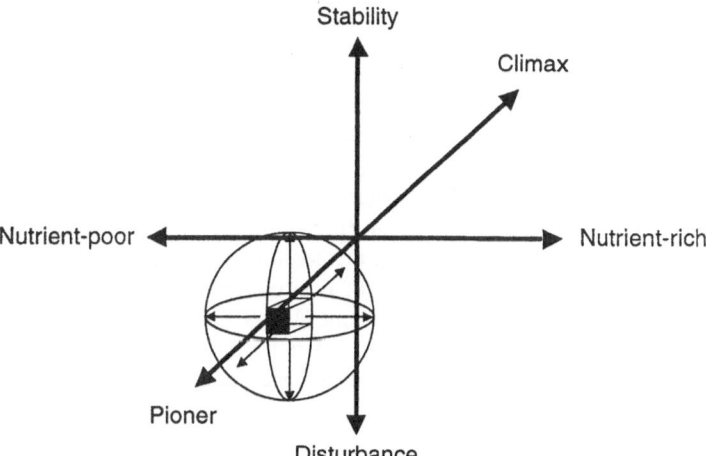

Figure 4.1 The box and the ball represent the ecosystem/plant community where the box is the basic state and the ball the allowed natural variation. The arrows, which are parallel to the axes, represent important vegetation determining gradients in which variation can take place. Changes beyond the limits of the ball could be due to non desired impacts, such as direct or indirect effects of GMHP, lack of suitable management or nitrogen deposition (see Section 4.5).

Basic state for non-target effects

If suitable, other parameters than those concerning the vegetation can be integrated in the determination of the basic state. Biotechnological acquired resistance, especially if broadspectred, against particular insects and diseases directly calls for investigation of the basic state of the community of related non-target species. GM plants with allelopathic properties may also pose significant threats towards natural vegetation (Kruse et al. 2000). This should be considered both in the risk assessment and in the design of monitoring programs (Kjær et al. 1999).

Baseline determination - an example

The baseline serves as a frame of reference against which all future changes to the ecosystem may be compared. We have tried to illustrate the problems involved in determining the baseline with a theoretical example, which can be used as a model for treatment of specific cases (Figure 4.2). The example uses changes in population size (or density) of organisms, such as plants, insects or birds, as a measure for effects which may have been directly or indirectly caused by the use of GMHP (e.g., from cultivation, through invasion, etc.), but the comments also apply to other types of effects. The changes in population size in areas not affected by GMHP generally show both year-to-year variation, which can be quite large, and a general long-term trend of increase, stability or decrease. Cyclical changes, e.g. from predator-prey interactions, can also occur. Both short-term and long-term trends may be caused by a number of different factors, as discussed above. The baseline, which represents the periodic trend, can be estimated differently depending on the period of observation and the level of year-to-year variation. In the example (Figure 4.2), a baseline determined from the first 7 year period will indicate a stable, unchanged state, but monitoring data from the entire 17 year period indicates a decreasing population size with a negative slope of the baseline. Therefore, when a possibly affected organism is monitored, a negative trend (Figure 4.2) may initially seem larger than after an extended period of monitoring. Similar situations will also be possible for detection of positive trends in population development. When representative control plots are used, the power of the analysis will increase, because data from single years can be compared and used for analysis. Any differences in the level of population size between control sites and monitored sites at the start should also be considered.

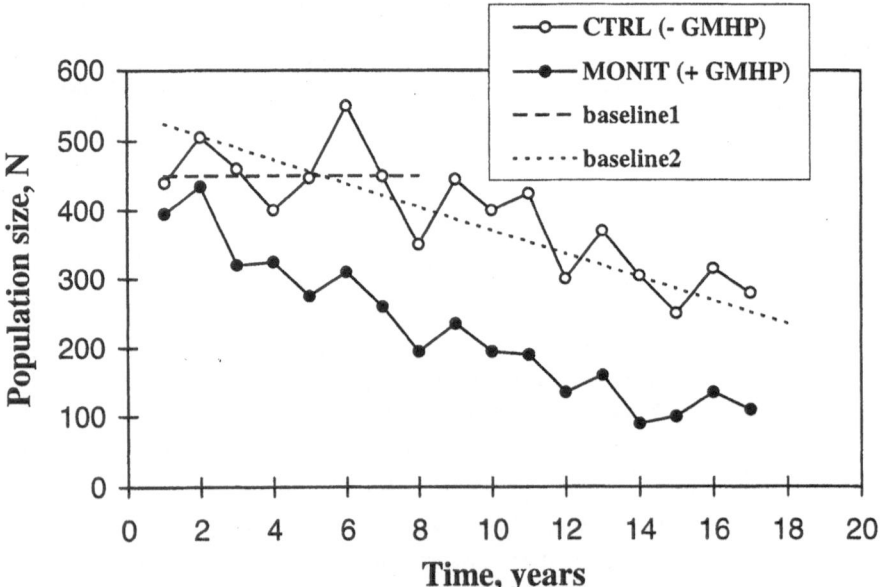

Figure 4.2 Theoretical example of baseline determination, effect detection and year-to-year variation based on different monitoring periods. The upper curve (open dot) represents control data used for baseline determination. Baseline 1 (dashed line) is based on a short-term period of monitoring and baseline 2 (dotted line) is based on a long-term period. The lower curve (dot) represents the data used for monitoring for trends and effects. See text for further information.

Baseline in a variable   Tyler and Olsson (1997) compared the flora of Skåne in Southern
environment   Sweden in 1938 and 1996 and found large changes in the number of

locations where species occurred due to environmental changes. Probably most ecosystems by nature change so fast that it is impossible to attain a static measure of their biotic and abiotic characteristics. Species often become extinct in one place and establish in another within a very limited span of years (Tyler & Olsson 1997). However, as illustrated above, a baseline need not to be static and general trends of increase or decrease can be included as well as variation and cyclic fluctuations. Such dynamics will be the rule rather than the exception, especially in cases of non-mature ecosystems. If the dynamics can be described mathematically, it is still possible to define a baseline, but it will take some effort and time, to attain the relevant data. Hence, alternatives have to be considered, such as the use of control plots or effect-distance relations. Here the first registration, the initial or the basic state of the ecosystem, can be used as starting point.

Surveys used for baseline   Surveys of regional or national changes to taxonomic groups, such as the flora, breeding birds and insects, may also be used as baselines for detection of local changes in species abundance (Harding 1991). Baseline surveys have for example been used for comparisons of trends in butterfly populations in single areas with regional trends in Britain (see Spellerberg 1991). The possible use of results available from large-scale surveys should also be considered when effects of GM crops are monitored at landscape scale (see Section 5.2). Procedures may include the use of multivariate statistics and geographical information systems (GIS) for analysis and interpretation of data. Furthermore, modelling for predictive purposes should be used.

5 Agricultural use and effects of GMHP

This chapter presents and discusses the specific problems and types of effects from GM crops that may occur in the cultivated fields and surroundings, and suggestions are made for surveillance and monitoring. General targets for monitoring of two different types of GM crops are presented (Section 5.1), and risk-aspects of large-scale farming are discussed in particular in relation to organic farming (Section 5.2).

5.1 Detection of adverse effects in cultivated fields

Effects on target organisms from the expression of the transgene, such as mortality of plant pathogens or insect pests, are generally not considered in the environmental risk assessment. Hence, monitoring of target effects is usually done primarily for agronomic purposes.

Monitoring non-target effects in the agro-environment

Monitoring possible non-target effects, such as food web effects to predatory insects, birds and mammals, becomes a major issue (see Section 7.4), especially when toxic compounds (e.g., Bt-toxin and protease inhibitors) and chemical products are involved. Any unforeseen effects to close relatives of target organisms or soil arthropods should also be considered (see Section 7.3). Furthermore, general impacts to the agro-environment caused by shifts in cultivation practice, such as changes in weed composition by the use of particular herbicides are also relevant for ecological monitoring.

Problems in monitoring non-resident organisms

For populations of mobile organisms, such as insects, mammals and birds, some species may not be permanently resident in the field habitat but only occasionally forage in the field. If these organisms are only exposed to the field environment part of the time, effects will be less likely to happen. This may also be the case for some organisms which are targets for insect-resistant GM crops. Effects to organisms which are sessile most of the time, such as plants, are easier to directly relate to the use of GM crops. The size of the field will also affect the possibilities for sampling organisms that have been negatively affected.

Effects of GM crops compared to effects from conventional crops

During risk assessment, the possible effects of the GMHP is generally compared to effects of the similar conventional plant. For farming, this includes treatment with pesticides, fertilisers and other management practices. It has been criticised that the GM crop system is being compared to conventional crop systems rather than organic or low input systems (Mayer 2000). Furthermore, there is widespread dissatisfaction with conventional farming in many countries. However, this is arguably the norm for comparisons based on overall importance and therefore also implemented in current regulation.

Targets for monitoring and surveillance

The targets for monitoring and surveillance in the farmland may be structured according to inserted trait, cultivation regime (conventional, organic and GMHP), proximity to GM plants, type of transgene dispersal and environmental effects. Two examples of this procedure, performed for herbicide-tolerant GM crops and insect-resistant GM crops, is outlined in Table 5.1 and 5.2, respectively. The differences in monitoring targets between the two types of crops mainly concern the type of effects, which needs to be monitored, while the monitoring targets for gene dispersal and plant establishment in the GM field are similar. The targets for cultivated fields (conventional and organic), which may be affected by the presence of GM fields, primarily include pollen dispersal, gene transfer and hybrid detection (see Chapter 6).

Table 5.1 Suggestions for minimum obligatory targets for specific monitoring and general surveillance of GM herbicide-tolerant crops in the farmland. Targets for surveillance are indicated by text in italics.

Monitoring target	Field with GMHP incl. adjacent surrounding areas	Areas outside GM field incl. conventional and organic fields
Gene dispersal (pollen, seeds) and GM plant establishment	Hybrids (crop dependent)	Pollen dispersal and hybrid detection (crop dependent)
	Volunteers, weedy crops (secondary establishment from seeds)	*Volunteers, weedy crops* (secondary establishment from seeds)
	Seed loss and dispersal with agricultural machines	*Seed spillage during transport*
Effects on organisms and processes	Weeds (density, diversity)	Not applied unless extensive GMHP dispersal is detected
	Insects, (density, diversity)	

Surveillance and monitoring of volunteers

Initial detection of volunteers in the fields may be included in the general surveillance of unspecified changes in the agro-environment done by the farmer or consultants. If persistent problems, such as weedy crops, tend to increase, this can require that specific monitoring is initiated. This will perhaps have to be followed by research to verify the causal factors, such as hybridisation, seed bank survival (Section 6.1) or lack of appropriate resistance management.

Seed loss during harvest may result in volunteers

Weedy crops and volunteers may arise through accidental seed loss. In seed crops, such as oilseed rape and cereals, and in seed production of other crops, a significant amount of the seeds may be lost due to natural shedding or spill from harvesting machinery. In swathed crops, overall losses ranging from 2.6 to 4.6 % and from 10.7 to 24.8 % has been reported for spring rape and winter rape, respectively (Price et al. 1996). When the rape plants were harvested directly, the losses were reduced. Seeds can also be transported between fields with farming machinery (Strykstra et al. 1997). Seed from GM crops which are lost

and incorporated in the seed bank may become a problem to crop rotation and cultivation of non-GM crops in the same fields. Management measures to reduce establishment of volunteer plants will be
required. This will often involve both mechanical cultivation measures
and herbicide treatment. The accidental dispersal of GM seeds by
sowing and harvesting equipment should be monitored by field sampling, and seed spillage during transport should be surveyed by
checking for weedy populations in adjacent fields and along roads.

Effects of GM herbicide-
resistant crops on weeds

The use of GM herbicide resistant crops will affect the weedy flora
through the changed use of herbicides, such as glyphosate and glufosinate. The effects on the flora depend on changes in dose, number of
treatments and the particular crop rotation regime (Madsen et al.
1997). The time for the treatment will also affect the chances of weed
reproduction. Impacts on plant species composition and density, biomass and reproduction (i.e., seed production) of selected species
should be monitored (Kjellsson & Simonsen 1994, see also Section
7.4). Knowledge on the weed composition before the GM crop is introduced in the particular field or region should be made available
(baseline information).

Table 5.2 Suggestions for minimum obligatory targets for specific monitoring and general surveillance of GM insect-resistant crops in the farmland.
Targets for surveillance are indicated by text in italics.

Monitoring target	Field with GM crop incl. adjacent surrounding areas	Areas outside GM field incl. conventional and organic fields
Gene dispersal (pollen, seeds) and GM plant establish-ment	Hybrids (crop dependent)	Pollen dispersal and hybrid detection (crop dependent)
	Volunteers, weedy crops (secondary establishment from seeds)	*Volunteers, weedy crops* (secondary establishment from seeds)
	Seed loss and dispersal with agricultural machines	*Seed spillage during trans-port*
Effects on organisms and processes	Insects, target and non-target (density, diversity)	Birds (density, clutch size)
	Birds (density, clutch size)	

Indirect effects of
pesticides on non-target
organisms

Indirect effects by changes in the use of pesticides by using GM crops
may be expected. Herbicides and insecticides have been reported to
reduce non-target populations of insect herbivores (Reddersen 1997)
and predatory birds, such as the skylark (Odderskær et al. 1997, Elmegaard & Andersen 1999). We suggest that monitoring should be done
primarily on insects and soil arthropods because birds, which are insect predators, are affected at a later stage. Methods for monitoring
soil arthropods and insects are described in Section 7.3 and 7.4, re-

spectively. Changes in population size and clutch size have been used for monitoring adverse effects to birds in general (see Section 7.4).

Reports of non-target effects on food webs

In large-scale field trials with insect-resistant GM cotton, the natural enemies of target insects were reduced in one year but showed no effects in another year with a new insertion of the transgene (Hardee & Bryan 1997). The difference could be explained either as a different plant/insertion response or as the result of different environmental conditions in the two years. It also illustrates the need for monitoring over an extended period of time.

Direct effects of insect-resistant crops

Information on the direct effects of insect-resistant crops on target organisms is primarily the concern of cultivation management and may therefore be provided by the agricultural monitoring or alternatively be included in the environmental monitoring program. The increased use of GM insecticidal crops with Bt-toxin may lead to a situation where target insects become adapted to the toxin. Different strategies to delay this adaptation have been reviewed by Gould (1998).

Additional information obtained from monitoring

Information on certain aspects (e.g., seed germination, survival, plant growth, seed production) in the life-cycle of the particular GM plant may be important to assess the ecological consequences of a release (Kjellsson & Simonsen 1994, Kjær et al. 1999). Some of this information may be obtained from monitoring or research done in small-scale field releases. Different GM lines of sugar beet (*Beta vulgaris*) with tolerance against virus infection and with herbicide-tolerance were monitored and showed different infection rates depending on soil conditions and an increasing winter survival with increasing temperature sum at different localities (Pohl-Orf et al. 1999, Schuphan 1999).

Monitoring field margins versus the centre

For plants, the spatial pattern usually shows larger variation across the field than along the direction of field treatment because the variation in seed bank densities is levelled by ploughing and other soil treatments (Rew & Cussans 1997). Furthermore, the density and diversity of weeds and insects are usually higher in the field margins than in the middle of the field, partly because pesticide treatment is less intense at the field margins. Monitoring in field margins may provide more information on potential changes due to the greater biodiversity and therefore the larger amount of possibly affected organisms. On the other hand, as a boundary zone to other habitats, the field margin is also subject to temporary invasion of organisms (e.g., some insects) that may not be well suited for monitoring of long-term effects.

Sampling procedures

Sample plots should be placed in rows along the field, both at the margins and towards the centre. For registration of weed densities, 0.1 to 0.5 m^2 large plots are usually suitable. The number of plots needed should be determined by a pre-trial to estimate the spatial variation in density and species content (power analysis should be employed, see Chapter 8). Normally from 20 to 50 plots in each row will be needed. The time for each census should relate to the activity of the organisms monitored. For weeds, data should be collected preferably two or three times per year depending on germination phenology and time of herbicide spraying. The time intervals for monitoring should be carefully considered to include information on year-to-year variation. Collecting information each year in the start period, e.g., during three to five years, and then at larger time intervals could be an effective procedure. For additional information see standard textbooks (e.g., Causton 1988, Kent & Coker 1993).

Crop rotation regime and the period of monitoring

The crop rotation regime will influence the onset and timing of any effects that may occur to the flora and fauna. In Denmark and Northern Europe, crops, such as oilseed rape, sugarbeet or potato, are normally cultivated only every third or fourth year in the same field (Madsen et al. 1997). This means that cultivation of a certain GM crop will require long-term monitoring, because effects, especially to the weedy flora, are not likely to happen after only one or two growing seasons. Furthermore, the soil seed bank tends to buffer any sudden changes to the composition of the flora (Cavers & Benoit 1989). The use of baseline information from before GM cultivation started and comparisons with non-GM control areas during the monitoring period will strengthen the possibilities for detection of effects (see BACI in Section 3.3).

Code of Practice and GAP for GM crop cultivation

The biotech companies usually design a Code of Practice for each specific GM product in agreement with the competent authorities. The aim is to provide guidance to the on-farm management of the product and reduce the risk of any accidental hazards. For GM products, instructions for Good Agricultural Practices, GAP, constitute the main part of the Code of Practice (Oppenheimer et al. 1996). The instructions usually include routines for: information of farm staff, recording of crop rotations, optimisation of management (e.g., pesticide use), seed storage and agricultural monitoring. The procedures for monitoring and management may concern detection and removal of boulters and volunteers, instructions for shallow cultivation, etc. Furthermore, the different general codes of practice which exist in EU-countries are to be followed, e.g., instruction in commercial use of herbicides, guides to seed certification, regulation of labelling, etc. The Code of Practice for GM products is communicated to the farmer, and technical consultants and distributors make sure that the content is understood. For each particular GM product, technical instructions will usually accompany the seed bags.

5.2 Effects and monitoring of GM crops used in large-scale farming

Extensive use of GM crops may have some major influence on product dependency for the individual farmer, which also affect farming in general. This includes the potential monopolisation of the market by single biotechnological products and companies and consumer acceptance of these products in general or for particular purposes (EC 1997). Furthermore, particular problems concerning seed and product purity may affect farmers involved in production of certified seeds and farmers involved in organic production.

EU regulation and the position of organic farmers

The current EU regulation for organic farming (EC 2092/91 and EC 1804/1999) exclude the use of GM crops in organic food production throughout the community. This is in agreement with the general opposition towards genetic engineering of food crops, which exists within organic farming. Some of the issues raised by organic farming organisations involve removal of right of choice for farmers and consumers, and violations of farmers rights and independence, besides fears of environmental impact and possible threats to human health (Young 2000).

Low levels of GM contamination may have to be accepted in organic products

If GM crops which produce fertile pollen are grown in the same region as organic crops of the same species, contamination by GM pollen cannot be totally excluded (see Section 6.1). For organic farming, a low level of GM contamination, perhaps 0.1 to 1 %, in certain organic crops, will have to be accepted by both farmers and consumers and implemented in current regulation, if GMHP are used in large-scale production. In any case, even if no traces of transgenes are detected in organic seed samples, there is still a low probability of contamination, due to the practical sampling restrictions.

Potential risks of GM crops to organic farming - seed contamination

Specific problems exist for organic farming to maintain seed purity and consumer product free of GM material. These mainly concern pollen dispersal, seed mixing and occurrence of volunteer plants in other crops (Moyes & Dale 1999). A possible further way of genetic contamination is through the use of seed from conventional crops which have received GM pollen. The use of conventionally grown seed is still permitted for organic crop production although it is being phased out (Young 2000).

GM pollen and organic beekeepers - detection difficult

Pollen collected by honeybees can become a constituent of the honey, which may cause difficulties to organic beekeepers whose bees are foraging on GM crops (Emberlin 2000). Honeybees can fly several km to forage, but usually forages closer than 1 km from the hive if suitable food sources are available (Waddington 1983, see also Section 6.1). Identification of GM pollen in the honey may be possible if the transgene is expressed in the pollen but will be difficult and costly. The activity of promoters, such as CaMv 35s and nos, has been suggested as targets for transgene detection in pollen for field release studies (Wilkinson et al. 1997). The detection level seems to be both variable and be depending on the particular plant species. Furthermore, a positive detection of contaminated honey will probably only be possible for high percentages of GM pollen depending on the sample sizes. The consequences for regulation, such as acceptance level for pollen content and isolation distance of GM crops from beehives, need to be considered.

Inducible gene control systems

A recent biotech development involves the use of inducible gene control systems, which enable plant genes and traits to be activated when needed. One particular aspect is the use of so called terminator genes where a promoter induces a lethal gene after seed formation, which will kill the seed but not other part of the plant (Crouch 1999). Consequently, the plant does not produce fertile seeds. In the 1998 patent (by Delta and Pine Land/USDA) the promoter is initially blocked, but plants and seed treated with a chemical inducer (tetracycline) releases the blockage and activates the gene. The technology was originally applied to a single plant strain, but has now been developed for production of hybrid seeds without the need of treatment with a chemical inducer (RAFI 2000).

Terminator technology - environmental concerns and farmers property rights

The terminator technology will reduce the risks of accidental seed dispersal and volunteer formation, and reduce pollen dispersal, if applied to production of male-sterile plants. Possible spread of terminator genes by cross-pollination with adjacent non-modified crops may, however, cause first generation hybrids which produce dead seeds (Crouch 1999). Furthermore, the potential dependency on biotech industry and reduced possibilities of growing second-generation seed has led to intense criticism from farmers organisations for reducing farmers property rights (RAFI 2000).

Scale of release influences the risks

Issues in risk assessment and monitoring will become more demanding when GM cultivation is scaled up from field trials to commercial release. This is because more organisms will be affected by the GM plants and will increase the probability of adverse and complex effects to happen. Especially for insect resistant GMHP which are able to produce hybrids with wild relatives, effects to non-target insects and the increased fitness of the modified plants could lead to greater uncertainty of risks (Stewart 1999). Furthermore, the problems to conventional seed production and organic farming caused by pollen and gene dispersal will increase considerably if GM crops such as oilseed rape cover extensive areas.

Weed populations and transgenes in the landscape

The landscape perspective should be considered when transgene escape and monitoring of hybrid weeds are planned. The spatial features of the landscape, including heterogeneity, dispersal distances, dormancy strategies and small-scale disturbances, influence the possibilities for coexistence of species (Lavorel et al. 1994). Introgression of transgenes into non-field populations of weeds may lead to adaptation to regulating factors such as herbivores and pathogens that are not active in field populations (Jordan 1999). Hence, weed populations in field margins and roadsides which become biotechnological resistant to insects will have increased fitness and recombinations with other traits of selective advantage will occur. This may lead to establishment of local metapopulations of certain species, which could become more resistant to management. The homogeneity of the agro-ecosystem may lead to rapid expansion of adapted weeds. Furthermore, hybridisation between weeds and crop species may occur at higher rates in isolated populations in non-cropped areas in the landscape than in field populations of weeds (Wilkinson et al. 1995). When monitoring transgene dispersal and establishment in field and marginal weed populations, the differences in genetic structure should be considered. The great variability of weed densities between years should also be taken into account (see Section 4.6).

Scale-up of GM crop cultivation increases the chances of detection of rare events

Small-scale field trials may not be designed to detect effects that occur at low probability, such as hybridisation with wild relatives near the field (OECD 1993). Furthermore, the efficiency of pest resistance in a new crop tends to be overestimated in small-scale field trials because the variation in ability of the target pest population to overcome resistance is not taken into account. When the cultivation of GM crops is scaled-up, the sensitivity of detection of rare events is increased - and even more so when the crop has been grown on a commercial scale for several years. Gene flow may also be expected to increase from small-scale trials to commercial cultivation (Klinger & Ellstrand 1999). Furthermore, when pollen populations increase this will result in an increase in the maximum dispersal distances that are recorded.

Surveys of general changes in pesticide use

At the landscape level, surveys covering distinct regions should be done to estimate any overall changes in pesticide use caused directly or indirectly by GM cultivation. Especially, when GM crops are used in mono-cultures covering extensive areas, such as the case in parts of North America, the use of special herbicides and insecticides could be affected (e.g., increased), but this has not happened yet (Fernandez-Cornejo & McBride 2000). Information of changed use may also be employed for predictions of effects to the environment by different cultivation scenarios comparing GM farming with conventional and organic farming (Madsen et al. 1997, Madsen et al. 1999).

6 Dispersal of GMHP and transgenes

This chapter covers background, methods and procedures for detection of dispersal and invasion of GM plants and gene flow and hybridisation involving the transgene. The GM crop plant can, depending on the species, disperse through pollination of other individuals, through seed dispersal, through survival in the soil seed bank or through vegetative fragmentation (Section 6.1). The inserted transgene can also spread to related species through cross-pollination and hybridisation (Section 6.2). Experience from monitoring trials and information on current regulation is provided, and different ways of detection and their limitations are discussed below. Indirect effects of GM crop use, such as weeds which become resistant to herbicides, and the consequences to management are also discussed.

6.1 Detection of GMHP dispersal and invasion

While some crop plants are largely self-pollinated, such as barley and peas, the majority relies on cross-pollination by insect pollinators or by pollen being carried by the wind (see Chapter 6 in Kjellsson et al. 1997 and Proctor et al. 1996). Methods for detection of pollen dispersal include (Kjellsson et al. 1997): pollen traps to determine concentration in air, trap plants and counts of pollen deposition and fertilisation, pollen viability tests, etc. Actual gene-transfer is best tested by using molecular markers present in the seeds. The different types of markers, which are available for detection of hybrids with weeds (see Section 6.2), such as DNA and herbicide-resistance markers, can also be used for detection of gene-transfer and crosses between crops in fields. The actual contamination rate, i.e., the percentage seed contaminated detected by trap plants, is generally much lower than contamination detected by simply measuring pollen concentration (Moyes & Dale 1999).

Pollination history of rape - interaction of many factors

Determination of pollen transfer and fertilisation often complicated by the flexible way in which flowers are pollinated. A short review of the pollination history of oilseed rape illustrates how different processes can interact. Self-pollination of rape can take place if flowers have not been pollinated by other means. Under field conditions, the majority of fertilised ovules normally result from self-pollination, although outcrossing in between 5 and 45 % of the crop has been reported (Metz et al. 1997, OECD 1997). Wind can carry pollen at least 2.5 km from a rape field to other fields. The concentration of pollen in the air decreases rapidly with distance from the source (Metz et al. 1997) and the effective contribution to pollination may be low. Social bees such as honeybees (*Apis mellifera*) and to a smaller extent bumblebees (*Bombus* spp.) and some solitary bees act as pollinators. Honeybees mainly collect nectar in rape flowers, but carry pollen attached to head and body between flowers. Honeybees show strong crop-constancy, and rape nectar may constitute the major food source in areas with abundant rape fields (Proctor et al. 1996). Flight distances of honeybees can be large, but normally not more than 1 km from the hive. In an EU-funded biosafety field project it was found that frequencies of transgene dispersal rapidly decreased with distance from the source (De Greef 1990). At four meters the out-crossing frequency was diminished to less than 0.1 %. Scheffler et al. (1993) in a study of pollen dispersal in transgenic oilseed rape, found a similar sharp decline of outcrossing frequency of 0.4 % at 3 m. and 0.02 % at 12 m. The variability involved in the above processes means that results from monitoring of pollen and gene flow generally need to be carefully interpreted.

Isolation distances for pollination

Minimum distances from neighbouring pollen sources have been set in the EU Seed Directives and have been applied to most of Europe. Isolation distances required for seed production in North America tend to be shorter than those in Europe are (Table 6.1). The isolation distances for field trials of GMHP normally confer to or exceed the minimum distances required for seed production. If these isolation distances are followed, it is likely that less than 1% of certified seeds of crops of beet, maize and rape will be contaminated by GM pollen (Table 6.1).

Table 6.1 Pollen dispersal and risk of contamination of some common crop plants, which are cultivated in NW-Europe and have been genetically modified. Minimum isolation distances used in production of certified seeds are shown. The seed contamination rates given are only approximate.

Crop species	Dispersal agent	Contamination rate (approx. range) [a]		Isolation distance [b]
		10 m	100 m	Europe/USA
Alfalfa (*Medicago sativa* ssp. *sativa*)	bees	-	-	(200) / 50.3
Beetroot, sugarbeet (*Beta vulgaris* ssp. *vulgaris*)	wind, (bees)	< 4 %	< 2 %	300-1000 / -
Maize, corn (*Zea mays*)	wind	0.05-0.2 %	(+)	200 / 201
Potato (*Solanum tuberosum* ssp. *tuberosum*)	bees	0.14 %	(+)	c
Rape (*Brassica napus* ssp. *napus*)	bees, wind	0.4-5 %	0.6-3 %	200-600 / 100.6

[a] : Information cited in Moyes & Dale (1999) if no other citation is given.
[b] : Distances in m for production of certified seeds (MAFF 1998, AOSCA 1999). [c] : No seed production, tubers are used for propagation.

Certified seeds of hybrid varieties of, e.g., rape are normally required to have a genetic purity of 99% (Moyes & Dale 1999). The efficiency of isolation distances for oilseed rape in the field has been determined by use of a herbicide-resistance transgene as a marker (Scheffler et al. 1995). High variability of gene flow between replicate plots and between individual plants in plots indicate that average rates of gene flow could underestimate actual gene transfer to parts of the field (Klinger & Ellstrand 1999).

GMHP field trials in the EU and isolation distances

Small-scale field trials with GMHP made in the EU usually conform to the above restricted isolation distances, e.g., 200 to 400 m for maize and 350 to 400 m's for rape. Usually the trials with, e.g., rape are surrounded by a 5 to 10 m broad zone of non-GM crop plants which will attract the majority of foreign pollinators. The plant material including seeds in the protective zone is usually destructed after the trial has terminated.

Herbicide tolerant crops and the evolution of resistant weeds

The use of a single weed management technology, such as biotechnological induced herbicide tolerance, may promote evolution of resistance in existing weeds (Duke 1999). Alternatively, overuse of one management strategy may lead to new weedy species become adapted to the agro-environment where other weeds have been removed. Monitoring of build up of resistance in single species and changes to the weed composition in the cultivated field will be needed.

Herbicide tolerance in oilseed rape and control of hybrids

In general, the inserted herbicide tolerance in a GM crop or hybrids will have a positive selective value only in agricultural environments, such as cultivated fields and surroundings, which are affected by spraying. Consequently, if the herbicide tolerance gene have a fitness cost outside the field, natural selection should make it less common. The fitness cost of, e.g., glufosinate tolerance, is however low. When transgenic herbicide tolerance of *Brassica napus* was introgressed into *B. rapa* in greenhouse trials, the glufosinate tolerance was not selected against (Snow & Jørgensen 1999). This indicates that the herbicide tolerant *B. rapa* could become persistent in field surroundings and make it more difficult to control.

Multiresistance and weed management

It can be a major problem to cultivation if weeds or weedy crops become resistant to a specific herbicide and even more so if they become resistant to two or more herbicides at the same time (i.e., multiresistance). Gene flow between *B. napus* cultivars and stacking of resistance genes in volunteer *B. napus* plants has been observed in Canada (Downey 1999). Because of the possibility for gene flow through pollen it was recommended that "growers should avoid sowing cultivars with different herbicide tolerances in the same or adjacent fields". A three-year monitoring study in France indicated that build up of multiple resistance to herbicides was a greater concern to GM farming of oilseed rape and sugar beet than outcrossing with wild relatives

(Champolivier et al. 1999). Monitoring of weedy GM oilseed rape for two years in England, has shown that the number of volunteers varied considerably from site to site depending on the persistance of the seeds in the soil (Norris et al. 1999). Hybrids with multiresistance to both glufosinate and glyphosate were detected, but were not considered to be more difficult to control than the single tolerant rape varieties. Management practices in these cases may involve the use of alternative herbicides such as 2,4 D and MCPA. The dependency of intensive herbicide use is, however, likely to increase if multiresistant crops and weeds become more common. This is likely to affect the density and diversity of plants and insects in the agro-environment in a negative way.

Transgene stacking in GMHP will increase

Monitoring and testing for the occurrence of multiresistant volunteers and weeds will be required when there are indications of problems in weed management. Different types of transgene stacking are highly desirable to GM plant breeding and likely to be of increasing importance in the coming years (Senior & Dale 1999, see also Section 1.2). At present, combinations of herbicide tolerance and insect resistance are much used, but combinations of three or four introduced traits have been reported for some of the EU release-trials with GMHP (http://biotech.jrc.it/gmo.htm).

Monitoring the spread of multiple transgenes will be difficult

The spread of different combinations of transgenes from many sources may become one of the major challenges to risk assessment and monitoring in the future. Undesirable combinations of transgenes could arise in wild relatives which could cause adverse effects to the environment. Multiple transgenes in hybrids will probably remain active for generations if they are closely linked (Chevre et al. 1998). The expression of the transgenes could however change in the new genetic backgrounds and cause unexpected effects to the environment. Monitoring for these effects will be difficult to target but crucial as a safety measure. If general surveillance of the farmland is implemented, specific monitoring procedures should be included for detection of stacked transgenes and their potential effects.

Techniques for reduction of transgene dispersal and hybrid formation

Different techniques have been developed which reduce the risk of trangene dispersal and hybrid formation. The use of a male-sterility system, which has been made primarily for the production of F1 hybrids, has been employed for GMHP such as oilseed rape, tomato and maize (Ruffio-Chable et al. 1993, Atanassova 1999). The method is environmentally unobjectionable, but it does allow external gene flow to enter the GM crop population (Darmency et al. 1995). Another type of technique which is currently being developed involves the use of terminator genes that prevent fertile seeds from being produced (see also Section 5.2). Although introgression could affect the fertility of wild relatives, there would be severe selection against the hybrids which produce sterile seeds. A negative aspects of both these techniques is that they reduce the possibilities for farmers in, e.g., developing countries of producing their own seed.

Crop seed banks, volunteers and monitoring

Some crop plants under cultivation can disperse and maintain viable seeds in the soil, e.g., beet and oilseed rape (Table 6.2). Other crops, especially large-seeded species such as maize, have very limited ability to persist in the soil. Differences between crops in phenology will influence seed dormancy. Seeds of the winter type of rape will germinate after a cold burial period when soils are turned and exposed to high temperatures (Crawley et al. 1993, Lopez-Granados & Lutman 1998). Seeds of the spring type are usually not dormant after ripening.

Table 6.2 Survival in seed bank and naturalisation of some common crop plants which are cultivated in NW-Europe and have been genetically modified. The range of longevity in soil seed bank (OECD 1997, Thompson et al. 1997) and the type of habitats where they occur as naturalised or as occasional are shown. If the crop species often occur as a weed in fields in NW-Europe this has been indicated.

Crop species	Seed bank year	Naturalisation, habitat types
Alfalfa (*Medicago sativa* ssp. *sativa*)	3 - 20	Roadsides, slopes, wasteland Crop normally not weedy
Beetroot, sugarbeet (*Beta vulgaris* ssp. *vulgaris*)	5 - 21 [a]	Roadsides, wasteland Crop weedy
Carrot (*Daucus carota* ssp. *sativus*)	1 - 35 [a]	Not naturalised Crop not weedy
Maize, corn (*Zea mays*)	0	Not naturalised Crop not weedy
Potato (*Solanum tuberosum* ssp. *tuberosum*)	7 - 10	Wasteland (rarely) Crop can be weedy (mainly by reproduction from tubers)
Rape (*Brassica napus* ssp. *napus*)	1 - 16	Roadsides, wasteland Crop weedy

[a] : The data refer to B. *vulgaris* ssp. *maritima* and the nominal species D. *carota*, respectively.

The seed survival also depend on the particular conditions in the field, including, e.g.: burial depth, seed mortality due to predation and pathogens, soil moisture and chemical conditions, use of fertilisers and herbicides (Leck et al. 1989). The survival of stray crop seeds in the soil can cause problems to conventional farming by establishment of volunteers and feral populations, which are particularly troublesome for GM crops. Traditionally, volunteers have been removed by spraying or by mechanical actions, such as removal of boulters of beet. Effective weed management and spraying can remedy some of the production problems caused by, e.g., weedy rape (Sweet et al. 1997), but at the cost of decreased plant diversity in the agro-ecosystem.

Persistent seed banks of weeds

The ability to maintain persistent seeds banks in soil together with annual life-cycle and high potential for seed production are characteristic for most common weeds (Cavers & Benoit 1989). The seed survival and the germination rate strongly influences weed population growth rates (Colbach & Debaeke 1998). While current weed populations are adapted to present management regimes, changes such as the use of herbicide-tolerant GM crops may affect seed production and seed bank of different species in different manners. Consequently, monitoring these changes to the seed input and resulting seed bank may be required.

Determination of soil seed bank content

The content of the seed bank can be determined from soil samples collected to, e.g., 20 cm depth, which are dried and/or cold treated and germinated in greenhouse trials (Kjellsson & Simonsen 1994). If data on seed production of weeds or crop seed spillage before soil treatment is required, sampling depths of 2 to 5 cm may be sufficient. Detection of rare species will need extensive sampling; hence these may perhaps better be identified from flowering plants in the field. Long-term monitoring for a period of minimum 5 to 10 years will normally be required.

Seed dormancy of hybrid crop plants

The ability of seed persistence in the soil may change when crop plants form hybrids with related species. Hybrid seeds from non-dormant oilseed rape (*Brassica napus*) and the dormant *B. campestris* seemed to be only slightly more dormant than rape (Landbo & Jørgensen 1997). Seeds from the first backcross generation of *B. napus* (female) x *B. campestris* (mail) showed increased dormancy and consequently should have a greater ability for survival in the seed bank. These processes would affect the chances for transgene dispersal and persistence of a GM rape crop in nature.

Monitoring protocols for agriculture and farmland

The concerns discussed above, about gene flow from GM crops to non-GM cultivars and to weedy relatives or seed persistence of volunteers, demand that specific monitoring objectives for the farmland are defined and procedures to deal with the needs are elaborated (Hill 1999). The main procedures that are needed to address these concerns are:

- Monitoring direct and indirect effects of the GM crop to the wild flora and fauna in the agro-environment. The type of effects dependent on inserted trait (e.g., herbicide resistance and insect-resistance) and target/ non-target organisms (see Chapter 5).

- Monitoring possible adverse effects of changes in cultivation practices by the use of GM crops compared to non-GM crops (e.g., caused by changes in herbicide use).

- Detection and monitoring the level of genetic contamination from GM crops to crops produced by organic farming (when acceptable threshold values have been defined).

- Monitoring the occurrence of multiple herbicide-resistant volunteer plants and hybrid weeds in the field and field surroundings.

- Surveillance and monitoring to detect possible adverse effects of transgene stacking in GM crops and hybrids with new genetic combinations to the agro-environment and surrounding habitats.

The general aspects of monitoring and surveillance of GMHP and the detection of effects on specific organism groups are covered in Chapters 3 and 7.

6.2 Detection of hybridisation and transgene dispersal

Crop-weed complexes with hybridisation between crop species, wild ancestors and intermediate weeds adapted to man-made habitats (Raamsdonk & van der Maesen 1996), make the analysis of gene flow complicated. Examples consist of: *Medicago, Vitis, Capsicum, Helianthus, Narcissus, Beta*, etc. For these complexes, gene flow will be difficult to control in distribution centres in the long term and demand extensive monitoring to follow.

Information on weedy relatives

The risk of hybridisation always has to be evaluated on a case-by-case basis and the monitoring procedures must include identification of the weedy relatives, which are locally present. Some basic information on potential hybrid formation with wild relatives is presented in Table 6.3, but specific relevant background information on hybrids should be available in assessment documents.

Transgene spread from crop to weed

There is a general lack of knowledge on the influence of fitness-related crop genes on population dynamics of wild relatives (Snow & Palma 1997). Traditionally, dispersal and introgression of wild genes into crop populations has been well studied and used during the evolution of new cultivars. Consequently, specific information on hybridisation between crop and weeds exist for many species (Table 6.3). For some crop species, cases of transgene dispersal to weedy species have been documented (Jørgensen et al. 1996, Jørgensen et al. 1998). Transfer of GM herbicide-resistance from rape (*Brassica napus*) to field mustard (*Brassica campestris*) was detected in field trials (Mikkelsen et al. 1996). Some types of genes introduced by modern biotechnology have similar effects to the environment as those of traditional plant breeding, and the consequences of introgression with wild relatives can be expected to be similar to those experienced from conventional hybrids (Dale 1994). Transgenes with novel traits (e.g., changed biochemical composition or stress tolerance) will need specific studies and monitoring to determine their chances for establishment in hybrids and the consequences to the particular habitats where they grow.

Table 6.3 Hybridisation between common crop plants and wild relatives. Hybrids, which have been identified to occur in field or natural habitats in NW-Europe or experimentally documented as likely, are shown. Literature references on hybridisation are given for each species. The list of preferred habitats is based on information in regional floras and biological literature (e.g., Grime et al. 1988).

Crop species	Hybrid formation with wild related species	Habitats of wild relatives	References
Alfalfa (*Medicago sativa ssp. sativa*)	*Medicago falcata* (hybrids are common in nature)	Grassy slopes, gravel pits, roadsides (prefers dry conditions)	Højland & Poulsen 1994, Rufener Al Mazyad & Ammann 1999
Beetroot, sugarbeet (*Beta vulgaris ssp. vulgaris*)	*Beta vulgaris ssp. maritima*	Seashores, stony ground and banks along the sea	Højland & Pedersen 1994, Bartsch & Schmidt 1997
Carrot (*Daucus carota ssp. sativus*)	*Daucus carota ssp. carota*	Grassland, wasteland, roadsides, slopes (prefers dry conditions)	Wijnheimer et al. 1989, Højland & Pedersen 1994
Maize (*Zea mays*)	No wild relatives in Europe	-	Kapteijns 1993
Potato (*Solanum tuberosum ssp. tuberosum*)	Wild relatives in Europe will not form hybrids with *S. tuberosum*	-	Højland & Poulsen 1994, OECD 1997
Rape (*Brassica napus ssp. napus*)	*Brassica campestris* *Brassica juncea* *Raphanus raphanistrum*	Arable fields	Højland & Poulsen 1994, Baranger et al. 1995, Darmency et al. 1995, Jørgensen et al. 1996, OECD 1997, Chevre et al. 1998, Jørgensen et al. 1998

Markers for hybrid detection

A number of different genetic markers can be used for detection of gene flow and hybridisation between crops and weedy relatives. For extensive reviews of marker types application and available detection methods see Kjellsson et al. (1997) and Parker et al. (1998). Marker genes are often used and inserted in the GMHP during the transformation process. The traditionally employed antibiotic and herbicide resistant markers (see below) can be objectionable for environmental reasons, but alternative methods are being developed (Harding & Harris 1997). The expression of the transgene or linked markers will in most cases to some extent depend on the environmental growing conditions for the GMHP. This must be considered when markers for monitoring are chosen. In a monitoring program for a commercially released GMHP it may be necessary to screen a large number of plants to detect gene flow and hybrids. This may limit the possibility to use advanced methodology; hence non-destructive simple techniques which have high reliability and are inexpensive, e.g., dot-tests of proteins, are needed (Simonsen 1999). Markers consisting of nuclear DNA (e.g., the genetic construct) are always preferable, as there is no effects of the environment on the expression, but they will also require elaborate facilities for analysis. Some examples of use of marker types for detection of transgene dispersal are presented below.

Herbicide tolerance used as marker

Some herbicide tolerant GM crops and hybrids are easily detected in field monitoring by applying a dot-test on plant leaves or spraying with the case-relevant herbicide, e.g., glyphosinate-ammonium (Scheffler et al. 1995). The use of a seed germination test with an assay for herbicide tolerance has also been suggested for monitoring transgenic plants, in particular a phosphinothricin tolerant oilseed rape (Pfeilstetter et al. 2000). The method gives rapid results and allows screening of a large number of seeds. An increasing number of GMHP have inserted two or more modified traits, and most often one of these includes some form of herbicide tolerance. If the herbicide tolerance genes do not segregate from other inserted traits during sexual reproduction, it may be used as a simple marker for the linked traits. However, stability is not always retained in more than a few generations, depending both on the fitness of the transgene and loss of expression of the transgene in hybrids (Metz et al. 1997).

Antibiotic resistance markers

The use of antibiotic resistance (e.g., kanamycin, ampicillin and hygromycin) as a marker linked to the genetic construct has earlier been used much for screening in transplant development, but is now being replaced by other safer methods due to the potential inherent hazards to human health.

Using DNA and RNA markers

Novel genes (construct DNA) inserted into the GMHP are in fact ideal markers to detect hybridisation and introgression into wild relatives of the crop (Ammann et al. 1997). The DNA-analysis can be performed using amplified restriction fragment polymorphism, AFLP, or other methods based on polymerase chain reaction, PCR (overview on methods in Kjellsson et al. 1997). Southern blotting has been intensively used for detection of genetically modified inserts. Because of the cost and amount of analysis required, these very precise methods can presently only be used for small-scale monitoring purposes. However, when analysis is automated or methods are reduced to dot tests they are ideally suitable (Simonsen 1999). The result of a PCR analysis (or any genetic analysis) needs to be carefully interpreted to avoid such examples as an incorrect report on pollen dispersal distances of GM potato up to 1000 m (Conner & Dale 1996).

Morphological markers

Qualitative morphological traits, such as flower colour (e.g., flavonoids) and leaf pattern, are biotechnologically available (Madsen & Poulsen 1997). When linked to the transgene construct, the morphological marker could be used for easy detection of the GMHP and hybrids. So far these techniques have not been applied to monitoring and detection. A method involving a fluorescent protein as in vivo marker, developed and tested in tobacco, has been suggested for use in monitoring GMHP (Stewart 1996). It is of course required that the marker is unobjectionable both to the environment and to public perception. Segregation between marker and transgene during sexual reproduction may give problems for practical applications.

Quantitative markers

Quantitative markers based on morphological measurements of taxonomic characters have traditionally been employed for detection of hybrids (often in combination with isozyme analysis) and have been suggested for GMHP as well (Mazyad 1997). However, many quantitative traits have multigenous inheritance and may easily become segregated during reproduction. Furthermore, variability of hybrid populations, possible segregation of linkage and construct and time consuming analysis make this method difficult to apply to practical monitoring.

GM crops with identical markers make field identification difficult

In the future, we must expect that different GMHP of the same species, which contain identical markers (e.g., herbicide tolerance), will be released in the same region. This will make a positive identification of the single GM product difficult or even impossible. The use of molecular methods, such as DNA or RNA markers, which constitute part of the genetic construct, may partly solve this problem. Another possibility to maintain a high detection level, is the use of a combination of markers of different types (Metz & Nap 1997). The ultimate goal will be the development of event specific detection assays for each single GMHP release.

Gene register for GM crops and transgenes

Certified DNA material from the specific GMHP is needed for reference when detection of dispersal and hybridisation is done. When different GM crops are cultivated, each transgene need to be identified separately in hybrids. In Germany, the Robert Koch Institute has planned the establishment of a gene register with, e.g., source information on transgene sequences, which can be used in large-scale monitoring and to reduce possible hazards of transgene interactions (Dreyer & Gill 1999). The revised EU-Directive 90/220/EEC will also include requirements for traceability of the genetic construct.

Methods for data analysis

A range of methods is available for data analysis and detection of transgene flow and hybridisation frequencies which can be based on the information on frequencies of molecular and other types of markers discussed above (Kjellsson et al. 1997), e.g.: F-statistics, genetic distance, genetic neighbourhood-size, paternity analysis and selfing and outcrossing rates. Furthermore, special computer programs are readily available for calculations. The problems involved in obtaining optimal sampling designs in studies of gene flow for different dispersal curves has been discussed by Klein & Laredo (1999).

7 Effect identification and detection

In this chapter, methods which can be applied for detection of adverse effects caused by GMHP are presented and discussed. This is, however, not an exhaustive account of the subject. In cases where commonly used scientific methods from other monitoring purposes are available, these are preferred and briefly described. Further information on individual methods can be found in the references.

Effect indicators and measurement parameters

Indicators of possible environmental effects constitute a broad range of animals and plants from different organism groups and ecosystems depending on the GMHP and the inserted traits (see Section 2.2). For a particular indicator species, a number of possible measurement parameters or fitness variables exist, including (Kjellsson & Simonsen 1994, Kjellsson et al. 1997): growth rate, biomass, reproductive effort, population rate of increase and genetic diversity.

In the present chapter, we will focus on changes in the population size and species composition as the main parameters for monitoring effects of GMHP. These parameters are relatively easy to analyse and strongly dependent on major changes to the environment.

7.1 Effect types, identification and assessment

Effects may be categorised in different types according to their mode of action and duration of time (see also Section 2.2 for specific definitions of effect types).

Direct and indirect effects

Direct effects relate to how the affected organisms or ecosystems are directly exposed to the GMHP or its dispersal in the environment. Indirect effects are caused by interactions of the GMHP with other organisms, e.g., through food chains. Usually, effects from the change in management practices caused by the GM crop are also considered as being indirect.

Long-term and delayed effects

Long-term effects are typically gradually changes to ecosystems taking place over time intervals of several years as a result of minor impacts. Typically they will only be detected after a considerable period of time, especially if they are unexpected. Effects, which occur after the termination of the impact, are normally called delayed. Both types represent some of the greatest challenges to the planning and establishing of effective monitoring and surveillance programs.

Short-term and immediate effects

Short-term effects occur relatively fast after the onset of the initiating event. In the case of an introduced GMHP, effect to the environment could occur immediately during the release or within three to five years after the release and would be expected to be of a well known nature. Examples include the immediate effects on target organisms, such as herbivores or pathogens caused by the inserted resistance genes. The chances of detection by monitoring would depend primarily on the level and extent of these effects.

Target and non-target effects

Target and non-target effects relate to the specificity of the inserted trait, or in other words, to the question whether the organisms are intended or not intended to be affected by the inserted trait. An example is the use of Bt-toxins in Maize aimed at the target species the corn-borer, but with potential non-target effects on other insects as well (e.g., the Monarch butterfly, see Losey et al. 1999).

Effects assessment and identification level

The identification of effects represents a kind of qualitative assessment. The extent of identified effects can be quantitatively assessed by use of detection methods that measure, e.g., how much a population is declining or increasing and also tell us and how fast these changes take place. The effects may be identified on:

- Organism level in the form of injuries and/or functional changes in selected species.
- Population level in the form of changes and/or extinction of populations.
- Species composition level (biodiversity level).
- Functional group level.
- Ecosystem level in the form of changed ecosystem function, i.e., food-web, food-chain changes and structural changes.

Time and space

Detection of effects on a larger scale in time and space demands for continuous monitoring on permanent plots and a distribution of monitoring sites representing the region in which effects are to be assessed.

Monitoring major consequences

In the present context, it will take us too far to treat methods related to all possible consequences and aspects of GMHP farming. Here we will primarily suggest methods for monitoring the major biological changes in ecosystem structure and function. Potential adverse effects and monitoring changes to the abiotic environment is mentioned only briefly. Methods for this purpose have been developed in connection with the integrated monitoring program. The aim has been detection of long-term abiotic changes in soil properties, amount of leaching, soil- and ground water chemistry, etc., due to acidification or excess nutrients. Methods recommended for monitoring soil- and ground water chemistry and soil properties are readily available (UBA 1996, UN-ECE 1998).

7.2 Assessment of effects on vegetation, ecosystem structure and function

The geomorphologic processes shaping the surface of the land area and the pedological properties of the soil are prerequisites for the vegetation, as well as they are important for soil living organisms. By offering space and food to other organisms, the vegetation is another important prerequisite for ecosystem structure and function.

Change of land use may affect the ecosystem

Shifts in land use normally induce changes in ecosystem structure and function. Soil-erosion and altered soil-properties, leaching and vegetation composition are examples of what may be expected in connection with the introduction of new agricultural practices. It should be noted that the introduction of new agricultural practices and new crop plants to all times have had unforeseen effects to the farmland. However, with the introduction of GMHP, any possible harm to the environment will not be acceptable to the public in general. Probably, the risk of unwanted effects is even higher for conventional plants, which are introduced in gardens or elsewhere, because the risks are not assessed beforehand. If any environmental harm is discovered, which is likely to be caused by an introduced plant, this has to be proven before counteractions can be implemented. Furthermore, regulation and jurisdiction will have to be decided upon and research will have to be carried out. By putting these aspects straight beforehand with the GMHP, a good standard is set, which both protects the environment and the rights of the farmers.

Stress-tolerant GM crops cultivated in marginal areas

The land use change associated with the introduction of GM plants is presently only in North America of a scale that may influence the agricultural landscape. On a longer time perspective and seen in the light of an increasing global population, areas not traditionally used for agriculture, such as saline soils and dry badlands, may be assessable for cultivation of stress-tolerant GM crops. This will perhaps of benefit to the requirements for food in the 3rd world, but can also lead to major ecosystem changes. These matters will have to be decided upon if and when the situation arises (see, e.g., Nielsen et al. 1998).

7.3 Detection of effects on soil organisms

Soil organisms constitute different functional types

Soil living organisms constitute a heterogeneous assembly of very different functional types of organisms. Most play a role in the decomposition of organic matter, which is of the highest importance for the function of terrestrial ecosystems. We have here selected a few groups which in most ecosystems appear to be the most important (Weaver et al. 1994, de Ruiter et al. 1996). This selection is necessary because it is impossible to monitor all groups. The functionally most important groups include nematodes, protozoa and microbes (bacteria and fungi). Protozoa are very difficult to sample and identify so they are suggested omitted in the context of GMHP monitoring. Estimation of bacterial diversity is often difficult because of methodological constraints and small sizes of organisms (Trevors 1998). Their activity can, however, be measured in different ways; hence for this group we suggest the use of entirely functional measures for monitoring. For measurements of bacterial diversity including genetic, species, and functional diversity, methods have been reviewed and proposed by Trevors (1998). Earthworms have been included below, because they are important for many ecosystems, e.g., for most agricultural soils. Furthermore, they constitute a well-known group of animals also in non-expert circles.

7.3.1 Microbiological activity

Decomposers

The main components of the decomposer system in soil are bacteria and fungi, which are responsible for the microbiological activity. Microbiological activity is a measure of the mineralisation of nutrients within the ecosystem. If the microbiological activity is influenced directly or indirectly by the GMHP (Donegan et al. 1995) this may result in a change in the amount of available nutrients in the ecosystem. Changes in the composition of populations of decomposers will not necessarily imply changes in the net mineralisation. The ecosystem function may remain unaffected even though the fungi and bacteria are affected.

Nitrification

A sensitive measure of the microbiological activity is the nitrite and nitrate formation by nitrification from ammonium or organic nitrogen compounds. Methods for measurement of gross- and net nitrification rates are available (Hart et al. 1994), and methods for soil sampling to use for microbiological analysis are described by Wollum (1994).

7.3.2 Soil animals

Monitoring choice for soil animals

Soil animals make up a diverse stock of animals ranging from microscopic microfauna (e.g., protists) to mesofauna, such as nematodes, springtails and other arthropods. The macrofauna of soil living animals include larger animals such as earthworms, voles, moles and gophers. Many different methods have been developed with the purpose of assessing the population size and species diversity of these animal groups.

Invertebrates as a functional group

The invertebrates as a group make up an ecologically important functional group. At least four aspects could be considered when choosing groups to use for monitoring effects of GMHP. Firstly, the ecological importance is important, because effects that confer risk to ecosystem function are important to detect and avoid. Secondly, the distribution should be taken into account, because the more common, widespread and homogenously distributed an animal group appear to be, the more suitable it is for monitoring. Thirdly it should be possible, with a reasonable effort, to identify selected species which are chosen for monitoring. Fourthly, well-described standard methods for sampling and identification exist.

Protozoa

Protozoa occur in all ecosystems of the world. They are ecologically important because of their impact on soil processes (Ingham 1994). From an ecosystem function point of view they are obviously a good choice for monitoring purposes related to soil. However, as already stated above we find them too difficult to sample and identify to be included in a monitoring program. However, when a GM potato with pest resistance from inserted lectins was tested in a field release, the only significant impacts on non-target soil organisms was a ca. 40% reduction in soil protozoan populations and ca. 10% in potential microbial activity (Griffiths et al. 2000). For further information concerning sampling, registration and identification of protozoa we have chosen to refer to Ingham (1994).

Nematode sampling

Nematodes are as good as ubiquitous to all ecosystems on earth, very important in most ecosystems and may by number constitute up to 90% of all multicellular animals in soil (Ingham 1994). They are heterogeneously distributed even over small areas and different species take up different parts of the space. This means that if one species is numerous in one spot, another species, which is common in the same field, may be rare in that particular spot. This patchy distribution is demanding when it comes to the sampling of nematodes. Soil samples should be taken in a pattern which represents the chosen site, and therefore numerous replicates will be necessary. A number of different methods has been developed for extraction of nematodes from the soil: wet sieving, funnel method, density centrifugation and elutriation with a semiautomatic extraction machine. For detailed information concerning these and other methods including sampling, extraction and identification, the reader is referred to Ingham (1994).

Soil arthropods

Soil arthropods role in soil ecosystems include transport and chemical and physical transformation of soil and substrate. However, the most important function is now recognised as their catalytic regulation of microbial activity (Moldenke 1994). Soil arthropods can be extracted from soil samples through physical methods or behaviour modification. Physical methods are well suited for extraction of meso-arthropods (0.5 – 2.5 cm) and macro-arthropods (>2.5 cm). Moldenke (1994) critically reviewed and described methods for sampling different sizes of arthropods from soil. These methods include destructive sampling of soil for biota, which is a method where the soil is physically broken and sieved for extraction of larger arthropodes. Much used is Pitfall trapping, by which species living on the surface are trapped in a funnel and sampled in a container with preservation liquid (e.g., ethylene-glycol) placed below. Extraction methods from soil cores includes passive and behaviour modification methods in which a soil sample is subjected to a moisture gradient. The soil arthropods migrate along the gradient and are collected in the end.

Springtails

Springtails (*Collembola*) are among the most abundant and widespread arthropods in soil and litter, occurring in densities up to 10^4 - 10^5 individuals m^{-2} (Hopkin 1997). They are present in most soils from very cold habitats in the arctic to very hot and dry habitats, and can play a vital role in the decomposition of organic material (Petersen & Luxton 1982, Hopkin 1997). A typical representative in Europe is the genus *Folsomia*. Members of this genus are present in both agricultural and natural habitats and *F. candida* is used for a standard ecotoxicological test (ISO 11267: 1999). Springtails are able to migrate from unsuitable environments and population size and structures are possible to determine. Hence it is likely that environmental changes in the soil from GMHP cultivation can be detected in springtail populations. Concerning methods for sampling and extraction the reader is referred to Hopkin (1997) and Moldenke (1994).

Earthworms

Earthworms are important in many ecosystems, especially in neutral soils in temperate regions with adequate precipitation to keep the soil moist. The sizes of earthworm populations vary over the season. Therefore, sampling has to be carried out at the same time each year in order to reduce the effects of seasonal variation. This has to be accounted for in the planning of monitoring programs. Both the sizes of the populations and the species composition should be monitored as they may vary as a result of environmental stress (e.g., caused by the GMHP).

Earthworm sampling

The sampling of earthworms for estimation of population size and species composition is further made difficult by the vertical and horizontal distribution pattern in soil. The field of methods for earthworm sampling for the above purposes is very large and very complicated (Edwards & Bohlen 1996). A sound conclusion may be that there is not one good adequate standard method applicable for this purpose. Combinations of methods are necessary to apply, if the main part of the earthworms is to be encountered through sampling.

A question that should be raised in connection with the monitoring of soil invertebrates is how the sampling influences the populations. The approach using permanent plot is problematic. The smaller animals having high population densities can be sampled in subplots representing larger permanent plots. Earthworms, however, are not that numerous and their sampling should be planned carefully in each case where effect monitoring is decided upon. Edwards and Bohlen (1996) present and discuss numerous methods for different purposes.

7.4 Detection of effects on aboveground organisms

For aboveground organisms, the monitoring has been limited to include higher plants, insects and birds. Only some of the relevant functional groups of insects need to be included in monitoring programs.
Food chains or food webs are ways to describe how the species in an
ecosystem are interrelated functionally. Ground organisms are not
strictly separated from soil organisms, but for the clarity of discussion
they are treated in separate sections of this book. General sampling
designs and methods for data analysis of terrestrial biodiversity in the
Nordic countries are available in From & Söderman (1997).

7.4.1 Detection of effects on plant species

Since vegetation is considered a major object for the monitoring programs, methodological questions on this subject were thoroughly
treated in Chapter 3 and 4. Some questions on changes to plant genetics and population structure are treated in Chapter 6 and mentioned
briefly below.

Effects on plants can be separated into genetical changes and changes
in population dynamics (e.g., size and age structure). Genetical change
is caused by direct transfer of genes from the GMHP to natural populations of the receiver species, transfer to related species through hybridisation and horizontal gene transfer between unrelated species.
Population size changes include increases and decreases in the number
of individuals of each species, but also relative changes in the population structure (i.e., frequency of different age classes or life-stages).

Genetic changes

Genetic changes in population can be detected by use of various genetic screening methods, such as PCR and AFLP with construction of
a genetic map (see further details in Kjellsson et al. 1997 and Section
6.2).

Population changes

Population size changes can be detected by use of frequency measures,
cover measures, biomass measures and by direct measures of population structure by, e.g., life-tables and models (Kjellsson & Simonsen
1994). Genetics may also be applied for estimation of effective population size (Kjellsson et al. 1997). They may also be assessed by use of
nature quality measures/models (Mark & Strandberg 1999).

7.4.2 Detection of effects on insects

Insect pollinators

Most crops are regularly visited by insects, which collect nectar or
pollen for food. If changes are made to the constituents of the food or
its availability, this may affect the pollinator species (e.g., the brood)
negatively. The most important pollinators of cultivated plants include

bees, bumblebees, and other insects such as butterflies, wasps and hoverflies (Kjellsson et al. 1997). Most pollinators also visit a range of wild plant species. Any decrease in population size would therefore negatively affect reproduction of the wild flora. Furthermore, adverse effects to ecosystem function could be the result, because many pollinators in their life cycles are associated with rare or decreasing elements of the environment like old woodland and stone fences. Besides being an important part of the ecosystem, they are also indicators of valuable nature or high nature quality (Dahl 1997, Nygaard et al. 1999b). Guidelines for an acute test for exposure to toxic compounds through pollen and nectar exist (Stute 1991) and is usually applied for honeybees (*Apis mellifera*). For each GM plant, tests for the relevant pollinator species must be made, and this should ideally include larval survival.

Insect herbivores and predators

Recommendations for methods concerning monitoring of aboveground insects are presented for butterflies, moths, beetles and pollinators (Southwood 1978, From & Söderman 1997). Although a certain method may be suitable for a given insect species or group, it is impossible to devise general methods that apply to all species. The biological behaviour of the insects, such as herbivorous or predacious foraging, should be taken into account in each case. Probably a combination of active and passive sampling techniques will be necessary.

7.4.3 Detection of effects on birds and mammals

Birds and mammals need to be monitored over extensive areas

Because of the high mobility of birds and mammals they will need to be monitored in large areas or in sessile phases, e.g., when nesting. It will be necessary in each case to identify the possible sensitive species. Their feeding biology, reproduction rate, breeding biology, age distribution, and population size should be compared to similar areas where the conventional crop is grown (control areas). Methods have to be decided in each case, probably an approach using large-scale investigations and controls will be necessary. A good assembly of methods for bird monitoring for different purposes can be found in Furness & Greenwood (1993). Birds have so far mainly been recommended in connection with environmental pollution, e.g., pesticides, heavy metals, euthrophication, etc., or larger changes in land use, such as forest management changes and draining and diking in marshy tidal areas. If the use of GMHP with time leads to extensive changes in agricultural land use, monitoring of birds should be seriously considered.

Toxic compounds and monitoring of birds and mammals

When birds and mammals come into contact with GM crops through consumption of plant tissue, they may be negatively affected if the plants contain new toxins or have higher levels of existing substances. Consequently, monitoring of changes to bird and mammal populations may be needed. This especially concerns cases where the metabolic content has been changed for production of proteins, enzymes and lipids (e.g.,"molecular farming"), pharmaceutical drugs or industrial compounds which could potentially affect larger herbivores. Indirect effects, caused by changed organism interactions, are perhaps not ideally monitored on birds and mammals which are top consumers. Organisms at the start of the food web, such as herbivore and predatory insects, will primarily be affected and should also primarily be monitored. Any changes to population sizes will also occur first at the start of the food web, and top consumers will be affected later. An exception to the above could be cases in which toxic compounds are accumulated from step-to-step in the food web. This would be a situation similar to the earlier experiences with pesticides such as DDT.

Species indicated for monitoring during risk assessment

When new GMHP products are marketed, the ecological risk assessment procedure should indicate particular relevant species or species groups for monitoring purpose. This must be based on ecological considerations of, e.g., predator-prey and plant-herbivore interactions or other food web considerations.

Mammals generally not suited for monitoring indirect effects of GMHP

After some consideration, we have reached the conclusion that mammals are not well suited for monitoring indirect effects of GMHP. Other vertebrates like reptiles, amphibians and fish may be in the same category as mammals, but the monitoring of these groups is beyond the scope of this book. Birds may be used as general indicators of effects from the changed cultivation practice, e.g., through their direct consumption of seeds and plant tissue.

7.5 Sampling designs and data analysis methods

Sampling designs are referred to or described in connection with the treatment of the different groups selected for monitoring purposes in this chapter and in Section 3.6. Procedures and method for detection of GMHP effects and dispersal in the farmland are described in Chapter 5 and 6. Additional sampling methods and data analysis methods can be found in the works referred to throughout this report, especially: Weaver et al. 1994, From & Söderman 1997, UN-ECE 1998, Lawesson 2000.

Statistical methods available for data-analysis

The statistical methods used for data-analysis should be decided already when the design of the monitoring plan is done. Different types variance analysis and regression have traditionally been much used (see, e.g., Sokal & Rohlf 1995), but methods involving cluster analysis, principal components analysis, data simulation and modelling are increasingly employed. For additional information on new statistical methods, the specialised literature should be consulted. For information on statistical analysis of genetic data, see section 6.2.

Power analysis used for sampling design	Statistical power analysis is especially useful in design of sampling programs. It is here used to estimate the number of samples and precision (i.e., sample variation) needed to detect differences between treatments, such as exposure in different areas, or to detect trends during time periods. Power analysis is fully described in Chapter 8.
Reduction of sampling errors	During sampling design, different ways to reduce measurement errors should be carefully considered. An optimal design will ideally reflect the true sample variation, e.g., in space, as is shown by the coefficient of variation, CV. Any reduction of variance will provide improved possibility to obtain statistical significance in effects detection.
Analysis of temporal trends	Analysis of temporal trends in GMHP abundance including one or more sites or habitat types is important for estimation of the potential for invasion and requirements for early warning. A method for analysis using regression is treated in Section 8.1. One additional method, which has a more general purpose for flexible sampling, is described below.
Sampling with partial replacement to detect trends in populations	In certain circumstances it may not be possible to identify exact locations and time of impacts caused by the release of GMHP. This may happen when the crop is used over extensive areas in crop rotation practices and shifting from farmer to farmer. In this case the sampling design has to be flexible to allow for detection of long-term trends on populations. The survey method of "sampling with partial replacement" (SPR) represents a well-confounded way to overcome some of the difficulties (Skalski 1990). A net of sampling stations has to be laid out covering the area, but with shifting sampling efforts. Sampling stations belong to a number of different rotational groups. During any one year, one group is replaced by a new series of sampling stations, and consequently sample overlap from year to year in the sampling program. Furthermore, new sampling stations may be added to allow for sampling in new locations with possible impacts. The composite estimator of the yearly population mean for the current year can be expressed as:

$$\overline{X}_0^{'} = Q\left(\overline{X}_{-1}^{'} + \overline{X}_{0,-1} - \overline{X}_{-1,0}\right) + (1-Q)\overline{X}_0$$

where Q is a weighing constant ($0<Q<1$), $\overline{X}_0$ is the estimated mean based on current year (0) samples, $\overline{X}_{0,-1}$ is the estimated mean for the current year based on samples common to year 0 and -1 (last year), $\overline{X}_{-1,0}$ is the estimated mean for the previous year (-1) based on samples only common to year 0 and -1, and $\overline{X}_{-1}^{'}$ is the composite estimator for the previous year, -1 (Rao & Graham 1964). Methods, which use substitution and partial replacement, are usually labour intensive and it may be difficult to reduce variation between sites.

8 Data analysis and evaluation

Planning and data analysis need to be integrated

The choice of pertinent methods for analysis of the different types of data from monitoring is crucial for an evaluation of the ecological effects. The statistical methods ideally need to be an integrated part of the monitoring plan already at the planning stage. This means that practical decisions on survey procedures, monitoring design, number of replicates, survey intervals, etc. all depend on the type of statistical tests which will be used to analyse a particular question. An important tool to help in achieving this goal is the use of statistical power analysis before survey start.

8.1 Statistical methods and tools

The different data analysis methods and statistical tests which are available to monitoring GMHP invasion and effects on organism groups and vegetation is presented in the respective chapters (See Sections 3.6, 6.2 and 7.4). This section will primarily concern general aspects of statistical power analysis with some examples illustrating the use in planning of GMHP data sampling. Baysian statistics, which involves probability estimates of different types of outcomes or models (Carlin & Louis 1996), are not included.

Statistical power analysis

In recent years the use of statistical power analysis has been increasingly applied to ecological studies and environmental monitoring, although the power to detect regional trends is seldom discussed (Urquhart et al. 1998). The power of a statistical test is the probability of getting a statistically significant result when monitoring a biologically existing effect. Statistically speaking, the probability of rejecting the null hypothesis when the alternative hypothesis is true (see Table 8.1).

Table 8.1 Decision errors of statistical tests in relation to power and significance. Power of a test is defined as 1- β, where β is the probability of making a type II error.

Null hypothesis, H_0	Test result	Decision/ Error	Probability
true	reject H_0	type I error	α
true	accept H_0	correct	$1 - \alpha$
false	reject H_0	correct	$1 - \beta$
false	accept H_0	type II error	β

Power increases with increasing effect size, sample size and significance level. Calculations of power will depend on the specific type of test and guidance to procedures and tabulated values are available to a range of standard tests (Cohen 1988). General statistical software may have some options for power analysis, but different stand-alone software cover many test types and are often easier to use (Thomas & Krebs 1996).

Effect size, ES

The term effect size, ES, is used to denote the size of change in the test parameter in relation to the expected value, incorporating variation. ES is the difference between the null and the alternative hypothesis. When the null hypothesis is true, ES = 0. Basically two types can be defined (Thomas 1997): 1. Standardised, dimensionless measures such as correlation coefficients or ES-indices (e.g., for ANOVA tests); 2. Raw measures such as difference between means or between slope of regression means. Raw effect sizes are generally easier to understand and interpret to test data. For the standardised type, it may be advisable to use the three categories of ES values suggested by Cohen (Cohen 1988): small (0.1), medium (0.25) and large (0.4). Exact definitions of effect sizes for different statistical tests can also be found in Cohen (1988).

Sensitivity analysis

When power calculations of assumed values are done, they should be followed by a sensitivity analysis. This means that a range of relevant values of main variables such as variance or effect size should be included in the calculations. The results, e.g., interactions between variables, can be shown in graphs or tables. This approach is also valuable to modelling (Section 8.2). Here it can be used to determine which variables have the strongest influence on results for a particular model.

Power values and significance level

A standard power value, $1-\beta = 0.80$ is often used when the number of samples needed are calculated to obtain a significance level, e.g., $\alpha < 0.05$, for a particular test. If the significance level is increased to e.g., 0.1, the power of the test increases. But this is only a way of reducing the probability of type II error at the expense of increasing the probability of type I error. In monitoring this may lead to an unnecessary alarm for adverse effects, so this procedure is questionable.

Recommended software for power analysis

Calculations of power for several types of tests (e.g., means tests, ANOVA and correlation) can be done using the statistical software PASS 6.0 (NCSS statistical software, http://www.ncss.com). Power estimates for population monitoring and detection of temporal trends can be made by the software MONITOR 6.2 (Gibbs, J.P., ftp://ftp.im.nbs.gov/pub/software/m) and TRENDS 1.0 using linear and exponential models (Gerrodette, T. 1993). Both software simulates linear and exponential growth, but MONITOR will only allow a maximum annual growth rate of 0.1. This may be sufficient for animal populations but not for plants. Thomas & Krebs (1996) have made a review of software that is currently available for statistical power analysis.

Examples of the use of power analysis in statistical planning of data sampling for GMHP monitoring are presented below.

Correlation analysis and power detection

Analysis of correlation can be used to detect association between monitoring variables and test for significance. Established association could lead to a proposed causal relationship. Observed correlation does not necessarily imply a simple relationship between variables - all sorts of interactions with environmental variables are possible (see Sokal & Rohlf 1995). However, correlation is useful as a simple measure of the possible interactions, which may require a closer analysis. Examples include: GMHP or transgene correlation to weedy species abundance, GMHP abundance and plant diversity, etc.

Power curves for correlation and number of samples

The use of power curves to estimate the needed number of samples for detection of significant correlation can be illustrated from Figure 8.1, which was produced by PASS. No variation of sample values is included so the analysis is only valid for singular but not for composite sample values. This will apply, e.g., for data collected at one site, but not necessarily for composite values representing several sites.

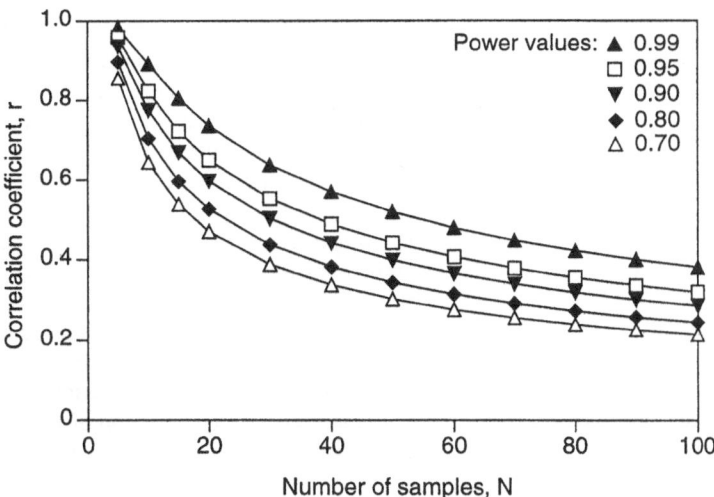

Figure 8.1 Analysis of power for detection of statistically significant deviation of a negative or a positive correlation coefficient, $|r| > 0$ at $\alpha < 0.05$. The power values represent a one-tailed test. The number of samples needed to detect significant deviations of $|r| \geq$ a specific value are shown for a number of power levels.

Example of use

As an example, it may be expected that a negative correlation existed between density of an invasive GMHP and a biodiversity measure for an organism group. Biodiversity could be number of plant or insect species, sum of frequency for a functional group of plants, etc. Measures are monitored per unit of area at one site. The question is: how many samples are needed to detect if a correlation coefficient, r, of a certain size is significantly different from zero (no correlation)? The power curves in Figure 8.1 can then be used to estimate the needed number of samples. The results are shown for a one-tailed test. To obtain an 80% probability of detecting a significant correlation of $r \leq$ -0.4 we need min. 38 samples, while a 99% detection probability would require a min. of 90 samples, which is probably unrealistic. If only high levels of negative correlation are to be detected ($r \leq 0.5$), 40 samples will ensure high power of detection ($1-\beta = 0.95$). If 20 samples are collected, a significant negative correlation of $r \leq$ -0.53 can be detected for power = 0.8.

General analysis of differences in effects and abundance

Traditional variance analysis (ANOVA) may be used for detection of changes in abundance or life cycle parameters of single species. Examples include direct and indirect effects on target and nontarget species, changes between habitats and during time. ANOVA methods were also employed by Crawley et al. (1993) for testing differences in demographic parameters between GM herbicide resistant oilseed rape and conventional rape in 12 sites during 3 years.

Power analysis for two-way ANOVA

An example of a power analysis for a two-way ANOVA of plant abundance in relation to number of sites and duration of treatment will be used to indicate relevant sampling numbers. We may want to detect if the number of times (e.g., years) a GM crop has been cultivated has any influence on weed abundance. Examples include herbicide resistant GM rape and potential increased occurrence of hybrid forms with *B. campestris*, which may have increased fitness under cultivation with herbicide treatment. The probability of hybrids with increased fitness will increase with period of use and fields with GM crops. Some assumptions need to be met: data normally distributed. However, abundance data (e.g., individuals m^{-2}) are often not normally distributed - a data transformation may be needed to fulfil this requirement including correction of zero values. Also, when sampling is done, spatial variation within a site is assumed to be nondirectional.

Detection of changes in abundance and between sites

The main object of the ANOVA is to detect changes in abundances with number of times of GMHP use and differences between sites (habitats). The null hypotheses tested by F-tests are that all effects for a term are zero. The power analysis should indicate the number of samples per site needed to detect a certain level of effect at power $\geq$ 0.8 and significance $\alpha = 0.05$. Small and medium effect sizes (ES = 0.1 and ES = 0.25) are employed because of the requirement for "early detection" of changes in abundance (see Cohen 1988).

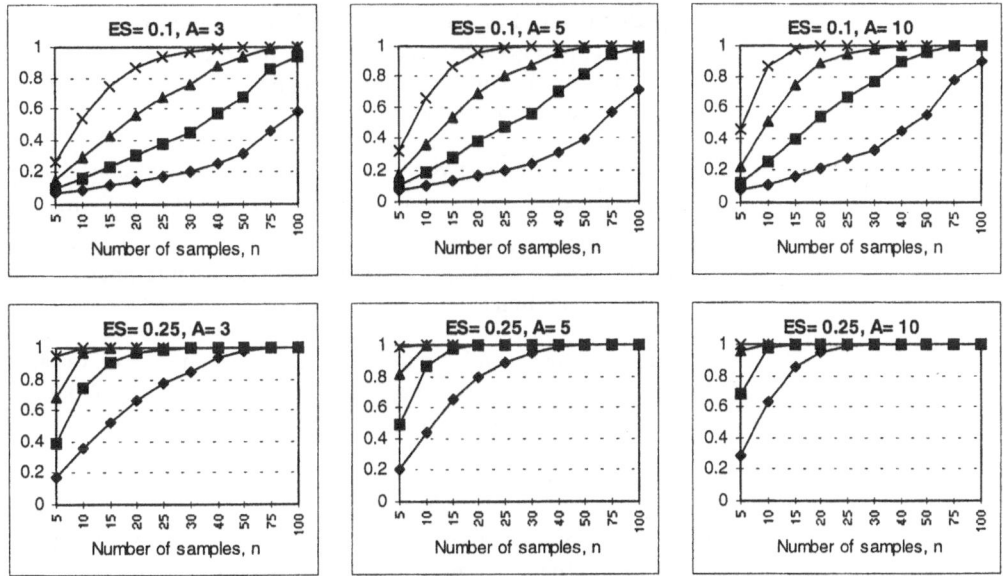

Figure 8.2 Power analysis for a two-way ANOVA of factor A (number of times, year) and B (number of sites). Separate graphs of power values (vertical axis) in relation to number of samples are shown for four levels of B, from top to bottom: 20 (x), 10 (triangles), 5 (squares) and 2 (polygons) sites. Results shown for two levels of effect sizes, low (upper row) and medium (lower row) based on simulations using PASS 6.0.

ANOVA power analysis results

The results of the power analysis (Figure 8.2) indicate that for medium effect size (ES = 0.25), changes in abundance can be detected after 3 to 10 years of use with > 80% power using 5 to 20 sites for monitoring with 15 samples per site and year. When a small effect size is chosen, we can only expect to detect changes at 80% power in 5 to 10 year periods with 5 to 20 sites and 50 samples from each site and period. If 5 different periods of GMHP cultivation are available for monitoring at 5 different sites, power for detection of medium ES approaches 1 with 25 samples, but power is only 0.47 when ES is small. To increase power ≥ 0.8 for ES = 0.1, at least 50 samples have to be collected.

Analysis of temporal trends

Early detection of environmental changes and trends in effect levels require careful consideration in sampling design and power analysis. Simple methods include the use of linear regression and tests of departure from zero slope of regression or any specified slope which is expected.

Analytic methods for detection of trends

Many different analytic methods have been applied to monitoring data of long-term trends in animal and plant populations, e.g., (Thomas 1996):

1. Linear-multiplicative models (e.g., linear route regression). The expected count, $E(L_{y+1})$, in year y+1 is a multiple of the count in year y, $E(C_y) = E(C_0)R^y$, where R is the trend. Models included are linear regression. An easily applied model, but it does not include changes in trends over time.

2. Polynomial-multiplicative models (e.g., poisson regression). Higher order effects are estimated by $E(C_y) = E(C_0)R_1^y R_2^{y^2} ... R_n^{y^n}$, for year y and the trend, R. Fluctuations between periods of years may be included by this model, but this may lead to weaker power of detecting a major trend.

3. Additive models involve removal of non-trend sources of variation (e.g., non-linear route-regression). No explicit assumptions are made about the shape of the trend.

4. Rank models trends (e.g., Spearman's rank correlation). Counts are ranked and the distribution of ranks over time is used to test if there is a tendency for an increasing or decreasing population. The direction of the trend, but not the size can be estimated by these non-parametric methods.

Simulation methods were employed to test the performance of three different methods on actual data from bird surveys (Thomas 1996). In general, route-regression methods were efficient in detecting trends but standard deviations were underestimated. Rank-trend analysis produced inaccurate results in some cases.

Autocorrelation between data

The problem of autocorrelation between data collected from the same area at different times can be resolved by special statistical methods, e.g., GLM-repeated measures. Methods for power analysis of these methods, however, are not available. In case of expected autocorrelation, power estimates made by similar available methods may be used for approximation. If variation is decreased by statistical correction for autocorrelation, these estimates will tend to be conservative.

Linear regression model for power analysis

A simple method of power analysis for population trend detection by linear and exponential regression has been suggested by Gerrodette (1987). In the linear model, abundance, A_i, is a function of i, an index of time:

$$A_i = A_1 (1 + r(i - 1)),$$

where A_1 is the initial abundance. The rate of increase, r, represents a change which is a constant fraction of A_1.

Exponential regression model for power analysis

In the exponential model, r is the finite fractional rate of change per time unit:

$$A_i = A_1(1+r)^{i-1}$$

The method for power analysis developed by Gerrodette involves five parameters: n, the number of samples; r, the rate of change; CV, the coefficient of sample variation; and α and β, the probabilities of type I and type II errors.

Rate of increase and overall change

For exponential growth, the value of r can be calculated for an overall fractional change in abundance, R, and a time period, n, from the equation:

$$r = (R+1)^{1/(n-1)} - 1$$

For detecting an overall increase of 300% after a period of 5 years monitoring, we calculate from the equation with R= 3.0 and n= 5 that r= 0.414. This value can then be used for detection of power and for comparison between the two types of rates. For the linear model, the equation is simply: r = R/(n-1). The comparable value in the example above then becomes, r= 0.75. Corresponding values of r have been tabulated for a wide range of overall increases (Table 8.2).

Expected rates of annual increase will be much smaller for perennial herbs and trees than for annual plants such as most GM crop species. This must be considered when the power analysis is made.

Table 8.2 Changes in rate of increase, r, in relation to the overall increase, R, and period of annual surveys. Values of r are tabulated for an exponential model.

years,	Fractional increase in period, R						
n	0.5	1.0	3.0	5.0	10.0	20.0	50.0
5	0.107	0.189	0.414	0.565	0.821	1.140	1.672
7	0.122	0.122	0.260	0.348	0.491	0.661	0.926
10	0.080	0.080	0.167	0.220	0.305	0.403	0.548

Power analysis of GMHP invasion trends

The method is here used to estimate power for detection of increase in an invasive population of a GMHP. In case the population is not seriously inhibited by competition (e.g., on bare ground and in open vegetation), exponential growth can be expected. If vegetation density is high and competition is severe, a linear model of population growth may be more appropriate. The natural expansion of the exponential model, the logistic growth curve, would perhaps have a wider application to field situations, but seems not to be implemented in current power tools. The calculations were performed by using the software TRENDS (Gerrodette 1993). It is assumed that CV is constant for A.

This was based on available information on plant population dynamics from a literature survey. If, however, $CV \in 1/\sqrt{A_i}$, as is sometimes assumed for quadrat analyses, power will increase. A correction for the use of the t-distribution for small sample sizes was employed (Gerrodette 1991). The level of sample variation in the simulations ranged from CV= 0.2 to 1. In the survey, it was found that for different annual plant species, CV ranged from 0.1 to 0.8, with means around 0.6. CV values for perennial species did not seem to differ much.

The results of the analysis clearly indicate the negative effect that a large CV has on the possibilities for early detection of an increasing GMHP population (Figure 8.3 and 8.4).

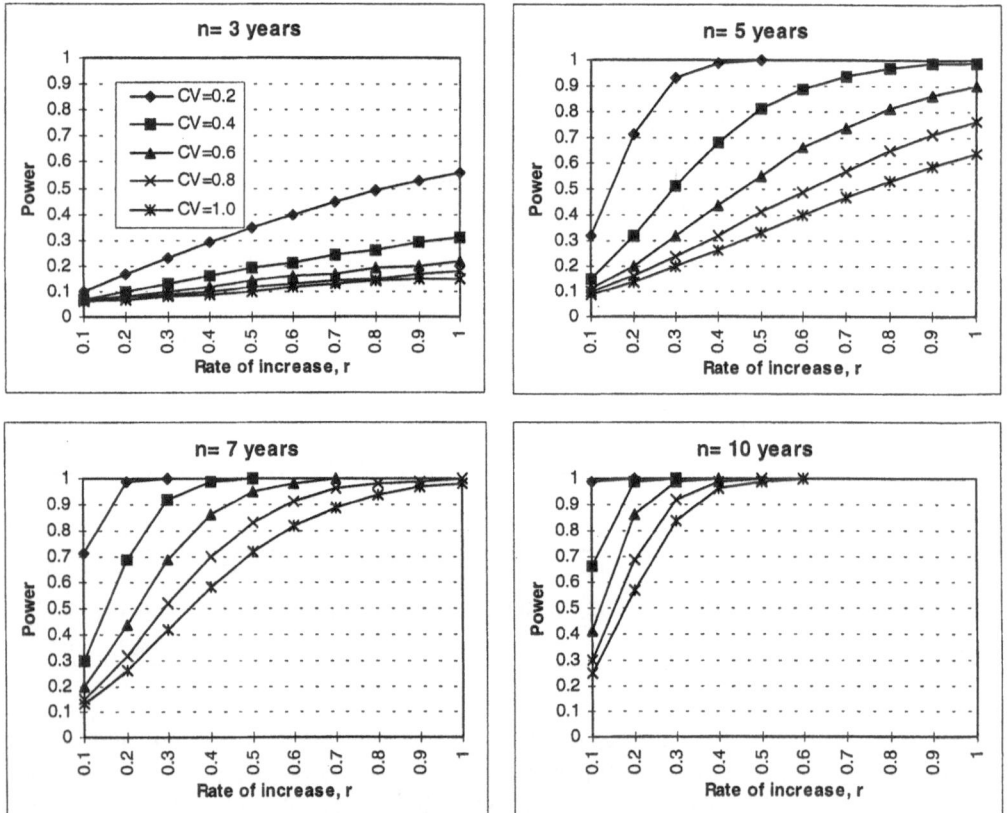

Figure 8.3. Power of exponential regression as a function of rate of increase, r, and CV for different number of years of annual surveys, n. Curves shown for an exponential model with α= 0.05.

The power curves suggests that an increasing trend is easier to detect for a population, which follows an exponential model, than one with linear growth (Figure 8.3 and 8.4). This seems to be a general conclusion (Gerrodette 1987). It is also evident that even for very high rates of increase, a sufficient power of trend detection requires more than 3 years of survey.

Exponential increase: detection of r

For a GMHP with an exponential increase, at least 5 years are needed to be able to detect a rate of increase, r= 0.5 for a variation level of CV≤ 0.4 (Figure 8.3). This corresponds to a total increase of approx. 300 % for the five-year period. If monitoring continues for 10 year, rates of increase down to r= 0.3 can be detected even when the level of variation is very high, CV= 1. Five years are often considered the minimum period needed for monitoring plant invasion in order to compensate from year-to-year fluctuations (e.g., Haber 1997).

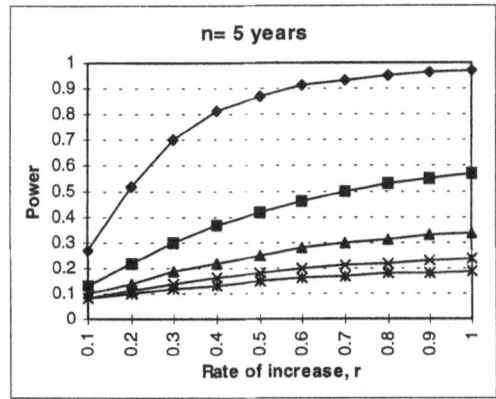 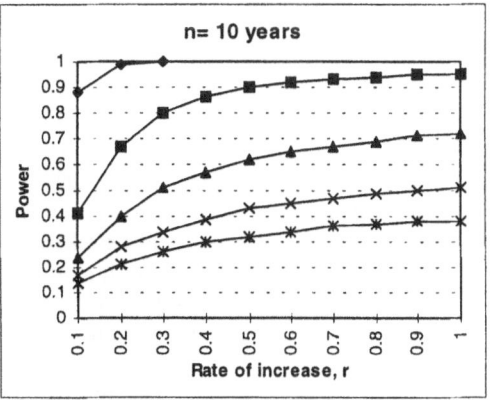

Figure 8.4. Power of linear regression as a function of rate of increase, r, and CV for 5 and 10 years of annual surveys, n. Curves shown for a linear model with α= 0.05. For identification of CV values, see Figure 8.3.

Linear increase: detection of r

For a GMHP with a linear change, an increase after 5 years can only be detected if CV≤ 0.2 and r ≥ 0.4, which corresponds to an overall increase of R= 1.6 Figure 8.4). After 10 years of monitoring, rates of increases, r ≥ 0.3 (≈ R= 1.2) can be detected down to a variation level of CV≤ 0.4.

Exponential increase and linear increase compared

The analysis shows that it is possible to detect rates of exponential increase, r ≥ 0.4 for even large variation level (CV ≤ 1.0) in a monitoring period of 8 years (Figure 8.5). For a plant following a linear increase model, detection of rates of increase, r ≥ 0.4 will only be possible if monitoring continues for 31 years at the same variation level. If however, CV ≤ 0.4 detection with 80% power will be possible after 9 years of monitoring (Figure 8.5).

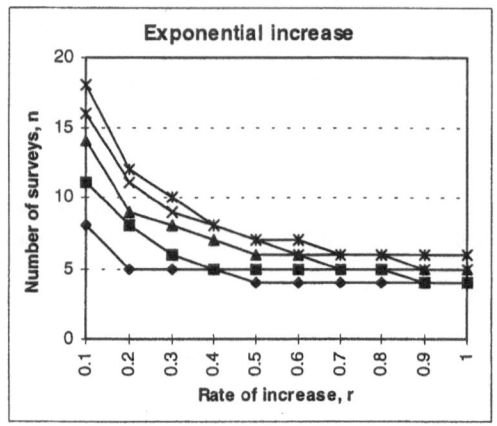 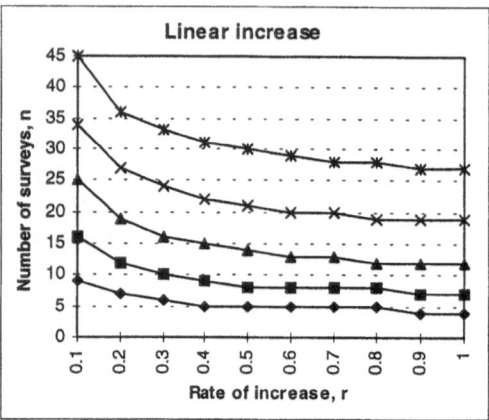

Figure 8.5 Minimum number of annual surveys (years) required to detect rates of increase, r, for different levels of variation. For identification of CV values, see Figure 8.3. Curves are shown for an exponential and a linear model both with β= 0.8 and α= 0.05.

Conclusions and restrictions of trend analysis

In general, detection of trends depends strongly on the number and precision of the samples (Gerrodette 1987). Consequently, measures should be taken which reduces variance of sample data (see Section 7.5). An important assumption of the trend analysis is that samples are assumed to be taken at regular intervals of time or distance. Another assumption of great concern for time series is that abundance estimates are independent. If there is positive autocorrelation, residual variance may be underestimated, which means that power of detecting a trend will be less than indicated by the analysis. In practical terms this means that analysis results shown above are minimum estimates of years, n, required and magnitude of trend, r, which can be detected; and maximum estimates of CV values which will permit detection (Gerrodette 1987). This is to some extent compensated by use of the t-distribution, which tend to decrease power estimates compared to the z-distribution suggested for larger number of subsamples (Gerrodette 1991).

Retrospective power analysis

When the results of a monitoring issue are analysed and evaluated, it is valuable to use power analysis to check for any consequences of deviations from the original assumptions. If for instance, sample variation is higher than expected, power of the test will decrease and the sampling program may need adjustment. This procedure is often called retrospective power analysis (Thomas 1997).

8.2 Use of modelling to assist monitoring

The use of modelling has been suggested as a general way to improve the effectiveness of monitoring trends (Edwards 1998). Statistical designs used to test for effects, as shown above, already depend on simple models, which typically include sample means and variation measures.

Population growth models
One type of model, which directly applies to GMHP success in different habitats, is a population growth model. The classical approach involves life tables or Leslie matrix with birth and mortality estimates for discrete life stages (see Burgman et al. 1993, Kjellsson & Simonsen 1994). A relatively large number of life-stage parameters have to be sampled for efficient estimation of the population growth rate.

Finite rate of increase used as a measure of GMHP invasion success
The finite rate of increase, λ, is a measure of the population increase per year that has been suggested as a measure of invasion success in natural communities (Crawley 1992). The finite rate of increase can be calculated as:

$$\lambda = e^{(b-d)}, \text{ and } N_t = N_0 e^{(b-d)t},$$

where b is the population birth rate, d is the population death rate, and N_0 and N_t is the number of plants in the population at the start and at time t, respectively. The factor "b-d" is also called the intrinsic rate of increase, r. λ represent the proportion by which the population changes by every time step, e.g., year. The threshold for a GMHP to successfully invade a community is that $\lambda > 1$ for the population. If λ is < 1, the population will decline and eventually become extinct. It has been argued that the finite rate of increase is a more appropriate response variable than the performance at any particular life stage because it combines effects of all different stages (Parker & Kareiva 1996). In annual species, λ can be estimated from the mean number of seeds produced per seed of the previous generation.

Assumptions and data requirements for λ
In practical terms, an estimation of λ in annual plants can be obtained from monitoring the number of reproductive plants in consecutive years. However, the estimation of λ is based on a number of assumptions of the growth model which may not always be valid, e.g.: no density-dependent effects, constant birth and death rates in time, no environmental effects are included, etc. (see Burgman et al. 1993). In an extensive three-year study of invasion success of oil-seed rape in different types of communities, the finite rate of increase was used to monitor invasion success (Crawley et al. 1993). A simulation analysis based on the results showed that the error of estimates of rate of increase decreased much stronger by increasing the number of sampling

years than by increasing the number of sites (Kareiva et al. 1996). This strongly suggests that monitoring efforts of less than three years duration produce poor estimates of the true λ value.

Plant population viability analysis and modelling

Plant populations of many species have been analysed for viability and the basic demographic data has been determined for modelling, determination of population growth rate, calculation of extinction probabilities, etc. (Menges 2000). In the majority of cases the length of the study did not exceed five years. The finite rate of increase, λ, was calculated in 82% of the studies. Such population viability analyses, PVA, can be useful as guidance in conservation and management (Menges 2000). The special problems, which exist for fitting population models to time series or making comparisons between time series, are discussed by Powell & Steele (1995). Temporal and spatial heterogeneity and stochasticity may make predictive modelling of small populations difficult (Gidding 1999).

Spatial models for population growth and gene flow

The spread of a GMHP into new areas clearly makes the spatial aspect of distribution and population dynamics important issues. A great number of different model procedures has been suggested, some involving patches, automata cells or diffusion rates (Lavorel et al. 1994, Cruywagen et al. 1996). The data requirements for assessment of spatial dispersal models have been evaluated and a discrepancy between the complexity of the model and the quality of the available data was found (Ruckelshaus et al. 1997). Model approaches to estimate gene flow of GM crops into conventional cultivars have also been used (Manasse 1992). Statistical considerations in analysis of space and time models of population dispersal and density dependence have been elaborated by Lele et al. (1998). A re-analysis of existing data on gene flow in selected crop species indicated a lack of appropriate data for detection of decay with distance in many trials (Gliddon 1999). Consequently, there is a need for analysis of the levels of detection possible with current sampling methods and monitoring protocols (i.e., power analysis).

Models for seed bank survival and seed dispersal

Survival of seeds in the soil seed bank and seed dispersal are two of the most important demographic stages that determine the success of plant populations to persist during unfavourable conditions and to invade new territories. Suggestions for population models involving seed bank extinction rates are numerous for both weeds and weedy crops (e.g., Fernandez-Quintanilla et al. 1986, Groenendael 1988, Jordan et al. 1995). Models for seed dispersal by wind depending on wind speed, seed weight and vegetation height are also available (e.g., Greene & Johnson 1989, Andersen 1991, Jongejans & Schippers 1999). Models have been developed for prediction of effects of the cropping system on transgene escape from GM rapeseed crops to rape volunteers (Colbach et al. 1999) and for prediction of gene escape of oilseed rape via the seed bank (Pekrun et al. 1999).

Detection of population change without long-term monitoring

A method to detect effects of environmental change on organism populations without the need for long-term monitoring data has been suggested (Doak & Morris 1999). The method relies on different stochastic demographic models and analysis of current population structure to make inferences about past and ongoing environmental change. It may apply especially to situations where monitoring information on the past is lacking.

Critique of modelling to predict plant invasion

Great scepticism concerning the use of models and field experiments to predict invasion of GMHP has been raised based on, e.g., analysis of historic data of weed invasion (Kareiva et al. 1996). However, it was argued, that the efficiency of sampling programs could be enhanced by use of modelling in design and sampling issues. Modelling and Monte Carlo simulation has been used to set confidence intervals for plant density estimation in a variable area transect (Engeman & Sugihara 1998).

8.3 Evaluation, information management and policy making

Analysis and evaluation of the results of a GMHP monitoring program should provide the relevant information needed for decision making concerning risk assessment, management and further monitoring (see Section 3.2). Detailed guidance to information analysis is beyond the scope of this book, but some general concepts, which may be used, are shortly discussed below. The ideas and principles, which are presented, have mainly been applied from the "WCMC Handbook on Biodiversity Information Management" (Reynolds 1998a). A framework for design and maintenance of ecological monitoring programs for environmental management has been proposed (Vos et al. 2000).

Requirements for good decision-making

Good decision-making depends on properly organised information, which provides clarity of complex issues and means of comparing potential solutions (Reynolds 1998b). This can be accomplished by:

- providing relevant information which is specifically tailored to the needs of decision-making (i.e., not too detailed),
- providing a range of options on which to base decisions,
- forecast consequences of each option and discourage the use of options with predictably adverse consequences, and
- adding information to a set of generally agreed facts for further discussion.

Information for, e.g., policy-makers and managers should be focused on operational goals and presented at the level of training (i.e., academic versus non-academic background). Furthermore, it should be available when and where it is needed. Information briefing should be supplied on changes of monitoring targets during time, e.g., the trend of representative indicators of ecosystem quality.

Network, information sharing and data access

Responsibilities and coordination between the different participants and information users in a monitoring network (e.g., industry, agriculture, research and regulation) should be attained. The benefits of information sharing and the common agreement on "need-to-know" should become evident to all participating bodies. It is important that principles for custodianship of data and procedures for exchange of information between bodies have been clarified (Reynolds 1998c). Agreement of access to data will typically involve questions on publication, intellectual property, acknowledgement of use, single versus multiple use, and cost.

Adjustment of monitoring procedures and decisions on management

For GMHP, new hazards may have appeared or hazards, which have been identified but deemed unimportant, may be found more significant than originally thought. This will require reassessment and further research. Adjustment of monitoring targets, methods and procedures must be considered and adjustment of input and structure of any models involved may be needed. Management measures to reduce adverse effects on ecosystems and organisms groups and halt further negative developments could be initiated if politically desired. However, most hazards that have occurred to ecological systems may be reduced to a certain level, but not completely removed once started.

Cost of management and changes in monitoring

Questions concerning the cost of management measures needed will in most cases depend on how urgent management is, the level of remediation accomplished, and the benefits involved (cost-benefit). Thus, the cost of, e.g., weeding plants in vulnerable areas, which can be grossly estimated as time x man-hours x wages, must be balanced by reduced impacts on other plants and organisms. The relative cost of change during the production of information will increase from the design phase of monitoring to development and use. Changing focus can therefore become costly and must be considered in relation to the benefits obtained by the adjustments.

Policy stakeholders and developments for the release and use of GMHP

The policy issues concerning the release of genetically modified organisms and higher plants are determined by a balance between economic, environmental and social goals (i.e., sustainable development). The main stakeholders which influence development in general are: governmental bodies (local and national administrators, international protocols and conventions), civic society (general public and non-profit organisations) and the private sector (industrial companies, trade bodies). The role of universities and applied research bodies vary depending on the interests involved, but are often partly independent of the three main stakeholders. Stakeholders from different levels and sectors of society participate in the development of policy goals by consensus or majority decisions taken at fora at different levels (Reynolds 1998c). The political development for use of GMHP is mainly influenced by (in no particular order): the biotech industry, national regulators and governments, international bodies such as the EU, trade and economic organisations (e.g., WTO and OECD), consumer groups and environmentalist organisations (e.g., Greenpeace), and academic research bodies (universities and applied research bodies). During the last 10-15 years there has been varying influence of the separate stakeholders and limited ability for consensus making between some interest groups.

Views on policy and concerns of different bodies influence the objectives of risk assessment and monitoring

The views on policy for biotech plant products taken by national governmental bodies and the general public have been different in Europe from that in North America. The main concerns have also differed between interest groups: regulation and risk assessment (national and international bodies), weed and resistance management (biotech industry and conventional farmers), concerns for natural habitats and organisms (ecologists and environmental organisations), market harmonisation (trade organisations), labelling of GM products (consumer groups), ethical concerns (general public). In Europe, the process of consensus and decisions on commercialisation of GM products has been drastically slowed down to what has been called "precautionary commercialisation" (Levidow et al. 1999). The policy decisions, which are taken, based on the influence of the different stakeholders, will ultimately also influence the objectives perceived as risks and influence the targets for both risk assessment and monitoring.

9   Conclusions and recommendations

From the analysis of the different aspects of monitoring GMHP fate and possible environmental effects, a number of conclusions were reached. Furthermore, suggestions for practical procedures in plot design, data sampling and analysis, have been presented. The major conclusions and recommendations, which we have reached are, summarised below.

Monitoring - a safeguard for environmental risk assessment - and early warning

The main purposes of monitoring and surveillance of GMHP are to assist the environmental risk assessment and risk management with relevant information for decision making. Risk assessment is often based on limited information on case-specific issues and assumptions are sometimes made from limited knowledge on the specific GM plant biology and organism interactions. The process is also confounded with a general lack of well-defined accept criteria. Therefore, monitoring is required to test the validity of the assumptions and detect any unexpected effects that may arise. An early detection of adverse effects caused by the GMHP should call for a reassessment and implementation of measures to reduce the consequences to the environment.

New biotech trends and monitoring requirements

New trends in biotech development (e.g., changed growth characteristics, altered stress tolerance, new chemical and industrial compounds) have significant consequences for both risk assessment and monitoring because new types of hazards may occur which will also influence the way they are detected. Hence, altered plant properties may have new effects on the environment, and new methods may be required for detection and monitoring.

Proposal for a cyclic monitoring scheme

A scheme is presented for the process of monitoring GMHP, based on the risk assessment, taking decisions on reassessment and management into consideration.

Monitoring split into three subprograms

The GMHP monitoring program should consist of three subprograms concerning: 1. Monitoring transgene dispersal, 2. Monitoring well-defined effects to the environment, and 3. General surveillance to detect unforeseen consequences and effects. Each of these three subprograms may be chosen alone or preferably used in combination depending on information from the ecological risk assessment. The more environmental concerns involved for the GMHP, the more should be included in the monitoring program.

Difficult to separate environmental changes from GMHP from other impacts

The changes to the environment caused by GMHP may be difficult to separate from other changes, such as those that originate from agricultural management practices. This will set high demands to the monitoring designs, which are similar in experimental set-up for both issues, and will also increase the requirements for the statistical analysis. Additionally, research may be required to test the relative importance of different factors (GMHP and others) for the registered environmental changes.

Monitoring primarily focused on direct effects - Surveillance focused on indirect effects

Monitoring should primarily focus on direct effects from the use of GMHP, because these effects are most likely to be adverse in the short term and occur at a level that can be detected. Indirect effects, such as food chain effects, effects to soil processes, etc., may take longer time to evolve and are less likely to be severe in the short term. The surveillance subprogram and general landscape monitoring should be able to detect indirect effects which are adverse. Exceptions could include effects from toxic compounds (e.g., Bt-toxins and allelotoxins), which are able to accumulate from step-to-step in food chains. But even in these cases, direct effects would be expected to take place first.

Planning of GMHP monitoring before release

Planning is an essential issue in all types of monitoring. In most cases the planning of a monitoring program is done only after hazards has occurred. However, the situation concerning GM-crops is more fortunate, because a monitoring plan is usually required before a commercial release is done. It is therefore possible to plan the details of case-specific monitoring prior to the release and optimise the chances for early detection of any adverse effects.

Power analysis should be used for planning and evaluation

It is strongly argued that statistical power analysis should always be employed when sampling designs for monitoring programs are planned and when the first results are evaluated. By doing this, optimal sample sizes and minimum monitoring periods can be determined for the required level of effects detection.

Reduction of sampling variability essential

In order to detect trends of changes in abundance of GMHP or indicator species, it is essential that the variation of estimated means is minimised during sampling. Hence the number and size of samples must be optimised to reduce sampling error and uncertainty.

Sample control data when possible

Control data should, if possible, be sampled from areas with similar biotic conditions and soil composition as the habitats where GMHP effects are suspected. These data will provide important reference information on the variation of indicators, which will increase the power of the effects detection.

Determination of baseline is essential for detection of changes

The basic state of the ecosystem, i.e., the baseline, must be determined before any changes caused by GMHP may be detected. This will include information on the structure, composition and variation of plant communities, insect populations and/or other indicator organisms depending on the specific case. The baseline is a prerequisite for the identification and explanation of changes observed by monitoring. In this way it serves as a frame of reference against which all future changes to the ecosystem may be compared.

Minimum 5 years needed to detect increases of GMHP

Power analysis of existing data and statistical considerations indicate that it will be impossible to detect any increase of a GMHP population in habitats where it is newly established during a period shorter than 5 years. In most cases, longer periods of monitoring will be required to detect significant changes in GM plant abundance.

Detection of adverse effects from GMHP use depend on scale of use

Detection of any effects caused by the use of GM crops is expected to require minimum 5 to 10 years of monitoring with at least 2 to 3 times of cultivation during crop rotation. The distribution pattern and scale of cultivation with GM crops in the farmland will strongly influence the chances of early detection of any adverse effects.

Selection of habitats for monitoring should be provided in the monitoring plan

The properties of the GMHP and the types of ecosystems, which may be affected, influence the choice of habitats selected for monitoring. Hence, when a new GMHP is approved for placing on the market, a selection of relevant monitoring habitats should be made, based on the case-specific ecological risk assessment. This selection, included in the monitoring plan, should be based both on plant properties and organisms, which are likely to be affected from the use of the GMHP. Selected agricultural fields where the GM crop is grown, and typical surrounding habitats should be monitored or at least surveyed. The surveillance program should therefore always include monitoring activities in habitats close to GMHP cultivation areas. In the dispersal and effects subprograms, the selection of monitored sites depends on the specific needs which are pointed out by the ecological risk assessment.

Selection of habitats based on invasibility

Some types of habitats, such as disturbed areas and some species-rich plant communities, are more exposed to invasion than others. Disturbed areas with low vegetation and high abundance of herbs and grasses are suitable for monitoring purposes. Firstly, because they are widely distributed and often found close to more intensively cultivated agricultural areas. Secondly, this vegetation type is interesting because roadsides, ditches and edges of fields often are the first to receive stray GMHP seeds and may therefore be used for early warning.

Monitoring targets for GM crops in the farmland

Suggestion for obligatory targets for monitoring and general surveillance in the farmland include hybrids, seed loss, volunteers and weedy crops in the GM fields. Monitoring in neighbouring areas (e.g., conventional and organic fields) should furthermore include detection of pollen dispersal and hybridisation from GM fields. The monitoring targets for effects detection depend on the case and inserted traits and should mainly be done in the GM fields and close surroundings. Monitoring changes in population density and species/genetic diversity of weeds, insects (target and non-target) and birds may be included.

Transgene stacking will increase the demands on monitoring

The increased use of multiple transgenes with different properties inserted into the same GMHP (transgene stacking) will make monitoring of gene flow and hybrid detection more complicated. Specific monitoring procedures should be developed to deal with this issue. It will also be relevant to monitor crops and weeds for acquired resistance to multiple herbicides as this may affect the use of herbicides.

Scale-up of GM cultivation influences effects detection

The scale-up of GM cultivation will increase the chances for detection of rare events, but also increases the demands on monitoring as more organisms and more complex interactions are involved.

Monitoring effects of large-scale use of GM crops

Additional issues involved in large-scale use of GM crops, which may require extensive monitoring, include the level of GM contamination of conventional and organic seeds and other plant products. Surveys should also be made to detect any major changes in pesticide consumption caused by large-scale use of GM crops.

Low level of GM contamination may have to be accepted in organic products

Even if current standards for crop isolation distances are increased, organic farmers and consumers may have to accept low levels of GM contamination, perhaps from 0.1 to 1%, in organic products, if GM crops are cultivated in large-scale in a region which also contains organic crops.

Modelling may assist risk assessment and monitoring

The use of population growth models and models for spatial distribution may provide means of assisting risk assessment in predicting gene flow and fate of invasive GMHP populations, although over-reliance of the outcome should be avoided. However, the efficiency of sampling programs for monitoring can be increased by the use of modelling.

Possibility of management and removal of invasive GMHP is limited

It will be possible to remove invasive GMHP or transgenes from vulnerable areas of limited size, but the removal may never be complete or able to cover extensive areas. Furthermore, it will probably also be costly to establish effective management and protection.

Consensus influence the possibilities for effective assessment and monitoring

It is crucial that consensus between major stakeholders regarding objectives for risk assessment of GMHP are at least partly achieved, if monitoring shall be able to assist in achieving politically determined goals.

10 References

Ammann, K., Felber, F., Jacot, Y., Jørgensen, R.B., Kjellsson, G., Olesen, J.M., Philipp, M., Rufener Al Mazyad, P., Schierup, M.H., Simonsen, V. (1997): Gene-flow/Hybridization. In: Kjellsson, G., Simonsen, V., Ammann, K. (eds.) Methods for risk assessment of transgenic plants. II. Pollination, gene-transfer and population impacts. Birkhäuser Verlag, Basel, 61-73.

Andersen, M. (1991): Mechanistic models for the seed shadows of wind-dispersed plants. Am. Nat. 137: 476-497.

Andersen, U.V. (1997): Monitoring invasive weeds at landscape level in Denmark. In: Brock, J. H., Wade, M., Pysek, P., Green, P. (eds.) Plant invasions: studies from North America and Europe. Backhuys Publishers, Leiden, 173-181.

AOSCA. (1999): Genetic and crop standards of the AOSCA. Association of Official Seed Certifying Agencies, Beltsville.

Atanassova, B. (1999): Functional male sterility (ps-2) in tomato (*Lycopersicon esculentum* Mill.) and its application in breeding and hybrid seed production. Euphytica 107: 13-21.

Bakke, T., Braathen, O.-A., Eilertsen, O., Myklebust, I. (1997): Quality assurance of field work. Nordic Council of Ministers, Copenhagen.

Baranger, A., Chevre, A.M., Eber, F., Renard, M. (1995): Effect of oilseed rape genotype on the spontaneous hybridization rate with a weedy species: an assessment of transgene dispersal. Theor. Appl. Genet. 91: 956-963.

Bartsch, D., Pohl-Orf, M. (1996): Ecological aspects of transgenic sugar beet: transfer and expression of herbicide resistance in hybrids with wild beets. Euphytica 91: 55-58.

Bartsch, D., Schmidt, M. (1997): Influence of sugar beet breeding on populations of *Beta vulgaris* ssp. *maritima* in Italy. J. Veg. Sci. 8: 81-84.

Bartsch, D., Sukopp, H., Sukopp, U. (1993): Introduction of plants with special regard to cultigens running wild. In: Wöhrmann, K., Tomiuk, J. (eds.) Transgenic Organisms - Risk assessment of Deliberate Release. Birkhäuser Verlag, Basel, 135-151.

Bichel Committee. (1999): Report from the main Committee - the committee to assess the overall consequences of phasing out the use of pesticides. Miljø- og Energiministeriet, Copenhagen.

Bijman, W.J. (1994): Potatoes and biotechnology: Technological development and social acceptance in the Netherlands No. 1.28. Agricultural Economics Research Institute (LEI-DLO), The Hague.

Blossey, B. (1999): Before, during and after: the need for long-term monitoring in invasive plant species management. Biol. Invas. 1: 301-311.

Bråkenhielm, S., Qinghong, L. (1995): Comparison of field methods in vegetation monitoring. Water, Air and Soil Pollution 79: 75-87.

Burgman, M.A., Ferson, S., Akcakaya, H.R. (1993): Risk assessment in conservation ecology. Chapman & Hall, London.

Burgman, M.A., Ferson, S., Akcakaya, H.R. (1993): Structured populations. In: Burgman, M. A., Ferson, S., Akcakaya, H. R. (eds.) Risk assessment in Conservation Biology. Chapman & Hall, London, 121-168.

Carlin, B.P., Louis, T.A. (1996): Bayes and empirical Bayes methods for data analysis. Chapman & Hall, London.

Causton, D.R. (1988): Introduction to vegetation analysis. Unwin Hyman, London.

Cavers, P.B., Benoit, D.L. (1989): Seed banks in arable land. In: Leck, M. A., Parker, V. T., Simpson, R. L. (eds.) Ecology of soil seed banks. Academic Press, San Diego, 309-328.

Champolivier, J., Gasquez, J., Messéan, A., Richard-Molard, M. (1999): Management of transgenic crops within the cropping system. In: Lutman, P. J. W. (ed.) BCPC Symposium Proceedings No. 72. Gene flow and agriculture - Relevance for transgenic crops. British Crop Protection Council, University of Keele, Staffordshire, 233-240.

Chen, J.M., Cihlar, J. (1995): Quantifying the effect of canopy architecture on optical measurements of leaf-area index using 2 gap size analysis-methods. IEEE Trans. Geosci. Remot. Sen. 33: 777-787.

Chevre, A.M., Eber, F., Baranger, A., Hureau, G., Barret, P., Picault, H., Renard, M. (1998): Characterization of backcross generations obtained under field conditions from oilseed rape-wild radish F1 interspecific hybrids: an assessment of transgene dispersal. Theor. Appl. Genet. 97: 90-98.

Cohen, J. (1988): Statistical power analysis for the behavioral sciences. Lawrence Erlbaum Associates, Hillsdale, New Jersey.

Colbach, N., Debaeke, P. (1998): Integrating crop management and crop rotation effects into models of weed population dynamics: a review. Weed Science 46: 717-728.

Colbach, N., Meynard, J.M., Clermont-Dauphin, C., Messéan, A. (1999): GeneSys: a model of the effects of cropping system on gene flow from transgenic rapeseed. In: Lutman, P. J. W. (ed.) BCPC Symposium Proceedings No. 72. Gene flow and agriculture - Relevance for transgenic crops. British Crop Protection Council, University of Keele, Staffordshire, 89-94.

Colwell, R.K. (1994): Potential ecological and evolutionary problems of introducing transgenic crops into the environment. In: Krattiger, A. F., Rosemarin, A. (eds.) Biosafety for sustainable agriculture - Sharing biotechnology regulatory experiences of the western hemisphere. ISAAA & SEI, Ithaca, 33-46.

Conner, A.J., Dale, P.J. (1996): Reconsideration of pollen dispersal data from field trials of transgenic potatoes. Theor. Appl. Genet. 92: 505-508.

Crawley, M. (1995): Long term ecological impacts of the release of genetically modified organisms. In: CEP (ed.) Pan-European conference on the potential long-term ecological impact of genetically modified organisms. Proceedings Strasbourg, 24-26 November 1993. Council of Europe Press, Strasbourg, 43-66.

Crawley, M.J. (1987): What makes a community invasible. In: Gray, A. J., Crawley, M. J., Edwards, P. J. (eds.) Colonization, succession and stability. Blackwell, Oxford, 429-453.

Crawley, M.J. (1992): The comparative ecology of transgenic and conventional crops. In: Casper, R., Landsmann, J. (eds.) Proc. 2nd Int. Symp. biosafety results of field tests of genetically modified plants and microorganisms. Biol. Bundesanstalt für Land- und Forstwirtschaft, Braunschweig, Germany, 43-52.

Crawley, M.J., Brown, S.L. (1995): Seed limitation and the dynamics of feral oilseed rape on the M25 motorway. Proc. R. Soc. Lond. B 259: 49-54.

Crawley, M.J., Hails, R.S., Rees, M., Kohn, D., Buxton, J. (1993): Ecology of transgenic oilseed rape in natural habitats. Nature 363: 620-623.

Crouch, M.L. (1999): How the terminator terminates: an explanation for the non-scientist of a remarkable patent for killing second generation seeds of crop plants. The Edmonds Institute, Washington.

Cruywagen, G.C., Kareiva, P., Lewis, M.A., Murray, J.D. (1996): Competition in a spatially heterogeneous environment: Modeling the risk of spread of a genetically engineered population. Theor. Popul. Biol. 49: 1-38.

Dahl, C.J., J. Larsen, H. Lawesson, J. Mark, S. Mogensen, B. Münier, B. Møller, P. Rune, F. Skriver, J. Søndergaard, M. Wind, P. (1997): Indikatorer for naturkvalitet. Arbejdssrapport Nr. 42. DMU, Rønde.

Dale, P.J. (1994): The impact of hybrids between genetically modified crop plants and their related species: General considerations. Mol. Ecol. 3: 31-36.

Danmarks Statistik. (1997): Statistical yearbook. Danmarks Statistik, Copenhagen.

Darmency, H., Fleury, A., Lefol, E. (1995): Effect of transgenic release on weed biodiversity: oilseed rape and wild radish. Brighton crop protection conference: weeds, No. 2. Proceedings of an international conference, Brighton, UK, 20-23 November 1995. British Crop Protection Council, Staffordshire, 433-438.

Darmency, H., Lefol, E., Fleury, A. (1998): Spontaneous hybridizations between oilseed rape and wild radish. Mol. Ecol. 7: 1467-1473.

de Greef, W. (1990): The release of transgenic plants into the environment: a review of the BAP projects. In: Economidis, I. (ed.) Biotechnology R&D in the EC. Catalogue of BAP achievements on risk assessment for the period 1985-1990, No. 1. Elsevier, Amsterdam, 19-22.

de Ruiter, P.C., Neutel, A.M., Moore, J.C. (1996): Energetics and stability in belowground food webs. In: Polis, G. A., Winemiller, K. O. (eds.) Food webs: Integration of patterns and dynamics. Chapman and Hall, New York, 201-210.

Doak, D.F., Morris, W. (1999): Detecting population-level consequences of ongoing environmental change without long-term monitoring. Ecology 80: 1537-1551.

Donegan, K.K., Palm, C.J., Fieland, V.J., Porteous, L.A., Ganio, L.M., Schaller, D.L., Bucao, L.Q., Seidler, R.J. (1995): Changes in levels, species and DNA fingerprints of soil microorganisms associated with cotton expressing the *Bacillus thuringiensis* var. *kurstaki* endotoxin. Appl. Soil Ecol. 2: 111-124.

Downey, R.K. (1999): Gene flow and rape - the Canadian experience. In: Lutman, P. J. W. (ed.) BCPC Symposium Proceedings No. 72. Gene flow and agriculture - Relevance for transgenic crops. British Crop Protection Council, University of Keele, Staffordshire, 109-116.

Dreyer, M., Gill, B. (1999): "Elite precaution" along with continued public opposition. Safety regulation of transgenic crops: completing the internal market? A study of the implementation of EC Directive 90/220. EC/DGXII/RTD biotechnology programme on the Ethical, Legal and Socioeconomic Aspects (ELSA), Bruxelle.

Duhn, K. (1994): Willows (*Salix* L. spp.) - Dispersal, establishment and interactions with the environment. The National Forest and Nature Agency, Copenhagen.

Duke, S. (1999): Weed management: implications of herbicide resistant crops. In: Traynor, P. L., Westwood, J. H. (eds.) Ecological effects of pest resistance genes in managed ecosystems. Information Systems for Biotechnology, http://www.isb.vt.edu, Bethesda, Maryland, USA.

Dunwell, J.M. (1999): Transgenic crops: the next generation, or an example of 2020 vision. Ann. Bot. 84: 269-277.

EC. (1997): European opinions on modern biotechnology. Eurobarometer 46.1. European Commission, Directorate General XII, Brussels.

Edwards, C.A., Bohlen, P.J. (1996): Biology and ecology of earthworms. Chapman and Hall, London.

Edwards, D. (1998): Issues and themes for natural resources trend and change detection. Ecol. Appl. 8: 323-325.

EEA. (1999): Genetically modified organisms. Environment in EU at the turn of the century, European Environment Agency, 245-261.

Ejrnæs, R. (1998): Structure and processes in temperate grassland vegetation. Ph.D.-thesis. National Environmental Research Institute, Rønde, Denmark.

Ekedahl, G. (1997): Guidance on quality assurance in environmental monitoring and assessment. Nordic Council of Ministers, Copenhagen.

Ellenberg, H. (1974): Zeigerwerte der Gefüsspflanzen Mitteleuropas. Scr. Geobot. 9: 1-97.

Ellenberg, H., Weber, H., Düll, R., Wirth, V., Werner, W., Paulissen, D. (1992): Zeigerwerte von Pflanzen in Mitteleuropa. Scr. Geobot. XVIII.

Elmegaard, N., Andersen, P.N. (1999): Food supply and breeding activity of skylarks in fields with different pesticide treatment. In: Adams, N. J., Slotow, R. H. (eds.) Proc. 22 Int. Ornithol. Congr. Durban. BirdLife South Africa, Johannesburg, 1058-1069.

Emberlin, J. (2000): Wind pollination. In: Allsopp, M., Parr, D. (eds.) GM on trial, Greenpeace, London, 5-12.

Engeman, R.M., Sugihara, R.T. (1998): Optimization of variable area transect sampling using Monte Carlo simulation. Ecology 79: 1425-1434.

Fernandez-Cornejo, J., McBride, W.D. (2000): Genetically engineered crops for pest management in U.S. agriculture: farm-level effects. U.S. Department of Agriculture, Washington.

Fernandez-Quintanilla, C., Navarrete, L., Andujar, J.L.G., Fernandez, A., Sanchez, M.J. (1986): Seedling recruitment and age-specific survivorship and reproduction in populations of *Avena sterilis* L. ssp. *ludoviciana* (Durieu) Nyman. J. Appl. Ecol. 23: 945-955.

From, S., Söderman, G., eds. (1997): Nature monitoring scheme - Guidelines to monitor terrestrial biodiversity in the Nordic countries Vol. 1997:16. The Nordic Council of Ministers, Copenhagen.

Furness, R.W., Greenwood, J.J.D., eds. (1993.): Birds as monitors of environmental change. Chapman & Hall, London.

Gerdemann-Knörck, M., Tegeder, M. (1997): Kompendium der für Freisetzungen relevanten Pflanzen; hier: *Brassicaceae, Beta vulgaris, Linum usitatissimum*. Umweltbundesamt, Berlin.

Gerrodette, T. (1987): A power analysis for detecting trends. Ecology 68: 1364-1372.

Gerrodette, T. (1991): Models for power detecting trends - a reply to Link and Hatfield. Ecology 72: 1889-1892.

Gidding, G.D. (1999): The role of modelling in risk assessment for the release of genetically engineered plants. In: Ammann, K., Jacot, Y., Simonsen, V., Kjellsson, G. (eds.) Methods for risk assessment of transgenic plants III. Ecological risks and prospects of transgenic plants, where do we go from here? A dialogue between biotech industry and science. Birkhäuser Verlag, Basel, 31-41.

Gliddon, C.J. (1999): Gene flow and risk assessment. In: Lutman, P. J. W. (ed.) BCPC Symposium Proceedings No. 72. Gene flow and agriculture - Relevance for transgenic crops. British Crop Protection Council, University of Keele, Staffordshire, 49-58.

Gould, F. (1998): Sustainability of transgenic insecticidal cultivars: integrating pest genetics and ecology. Ann. Rev. Entomol. 43: 701-726.

Greene, D.F., Johnson, E.A. (1989): A model of wind dispersal of winged or plumed seeds. Ecology 70: 339-347.

Griffiths, B.S., Geoghegan, I.E., Robertson, W.M. (2000): Testing genetically engineered potato, producing the lectins GNA and Con A, on non-target soil organisms and processes. J. Appl. Ecol. 37: 159-170.

Grime, J.P., Hodgson, J.G., Hunt, R. (1988): Comparative plant ecology. Hyman, London.

Groenendael, J.M. van (1988): Patchy distribution of weeds and some implications for modeling population dynamics: a short literature review. Weed Res. 28: 437-441.

Haber, E. (1997): Guide to monitoring exotic and invasive plants. http://www.cciw.ca/eman-temp.

Hardee, D.D., Bryan, W.W. (1997): Influence of *Bacillus thuringiensis*- transgenic and nectariless cotton on insect populations with emphasis on the tarnished plant bug (*Heteroptera: Miridae*). J. Econ. Entomol. 90: 663-668.

Harding, K., Harris, P.S. (1997): Risk assessment of the release of genetically modified plants: a review. Agro Food Industry Hi-Tech 8: 8-13.

Harding, P.T. (1991): National species distribution surveys. In: Goldsmith, B. (ed.) Monitoring for conservation and ecology. Chapman and Hall, London, 133-154.

Hart, S.C., Stark, J.M., Davidson, E.A., Firesone, M.K. (1994): Nitrogen mineralisation, immobilisation and nitrification. In: Weaver, R. W., Angle, S., Bottomley, P., Bezdieck, D., Smith, S., Tabatabai, A., Wollum, A. (eds.) Methods of soil analysis. Part 2. Microbiological and biochemical properties. Soil Science Society of America, Madison, Wisconsin, 985-1018.

Heywood, V.H. (1989): Patterns, extents and modes of invasions by terrestrial plants. In: Drake, J. A., Mooney, H. A., di Castri, F., Groves, R. H., Kruger, F. J., Rejmanek, M., Williamson, M. (eds.) Biological invasions: a global perspective. John Wiley & Sons, Chichester, 31-60.

Hilbeck, A., Baumgartner, M., Fried, P.M., Bigler, F. (1998): Effects of transgenic *Bacillus thuringiensis* corn-fed prey on mortality and development time of immature *Chrysoperla carnea* (Neuoptera: Chrysopidae). Environ. Entomol. 27: 480-487.

Hilbeck, A., Meier, M.S., Raps, A. (2000): Review on non-target organisms and Bt-plants. EcoStrat GmbH, Zurich.

Hill, J.E. (1999): Concerns about gene flow and the implications for the development of monitoring protocols. In: Lutman, P. J. W. (ed.) BCPC Symposium Proceedings No. 72. Gene flow and agriculture - Relevance for transgenic crops. British Crop Protection Council, University of Keele, Staffordshire, 217-224.

Hobbs, R.J. (1989): The nature and effects of disturbance relative to invasions. In: Drake, J. A., Mooney, H. A., di Castri, F., Groves, R. H., Kruger, F. J., Rejmanek, M., Williamson, M. (eds.) Biological Invasions: a global perspective. John Wiley & Sons, Chichester, 389-405.

Hobbs, R.J., Humphries, S.E. (1995): An integrated approach to the ecology and management of plant invasions. Conserv. Biol. 9: 761-770.

Hopkin, S.P. (1997): Biology of springtails. Oxford University Press, Oxford.

Højland, J.G., Pedersen, S. (1994a): Carrot (*Daucus carota* L. ssp. *sativus* (Hoffm.) Arcangeli) - Dispersal, establishment and interactions with the environment. The National Forest and Nature Agency, The Environmental Protection Agency, Ministry of the Environment, Copenhagen.

Højland, J.G., Pedersen, S. (1994b): Sugar beet, beetroot and fodder beet (*Beta vulgaris* L. ssp. *vulgaris*). Dispersal, establishment and interactions with the environment. The National Forest and Nature Agency, The Environmental Protection Agency, Ministry of the Environment, Copenhagen.

Højland, J.G., Poulsen, G.S. (1994): Five cultivated plant species: *Brassica napus* L. ssp. *napus* (Rape), *Medicago sativa* L. ssp. *sativa* (Lucerne/Alfalfa), *Pisum sativum* L. ssp. *sativum* (Pea), *Populus* L. spp. (Poplars), *Solanum tuberosum* L. ssp. *tuberosum* (Potato) - Dispersal, establishment and interactions with the environment. The National Forest and Nature Agency, The Environmental Protection Agency, Ministry of the Environment, Copenhagen.

Ingham, E.R. (1994): Protozoa. In: Weaver, R. W., Angle, S., Bottomley, P., Bezdieck, D., Smith, S., Tabatabai, A., Wollum, A. (eds.) Methods of soil analysis. Part 2. Microbiological and biochemical properties. Soil Science Society of America, Madison, Wisconsin, 491-517.

Ingham, R.E. (1994): Nematodes. In: Weaver, R. W., Angle, S., Bottomley, P., Bezdieck, D., Smith, S., Tabatabai, A., and Wollum, A. (eds.) Methods of soil analysis. Part 2. Microbiological and biochemical properties. Soil Science Society of America, Madison, Wisconsin, 459-491.

Jacot, Y., Ammann, K. (1999): Gene flow between selected swiss crops and related weeds: risk assessment for the field releases of GMO's in Switzerland. In: Ammann, K., Jacot, Y., Simonsen, V., Kjellsson, G. (eds.) Methods for risk assessment of transgenic plants. III. Ecological risks and prospects of transgenic plants, where do we go from here? A dialogue between biotech industry and science. Birkhäuser Verlag, Basel, 99-108.

James, C. (1997): Global status of commercialized transgenic crops in 1997. ISAAA Briefs, No. 5. ISAAA, New York.

James, C. (1999): Global status of commercialized transgenic crops: 1999. ISAAA Briefs, No. 12: Preview, ISAAA, New York.

Jensen, K., Christensen, M.K. (1994): Invasion af Vår-raps og Foder-Lucerne i naturlige plantesamfund. Københavns Universitet, Botanisk Institut, Økologisk Afdeling, København.

Johnsen, I., Ro-Poulsen, H., Søchting, U., Mortensen, L. (1991): Gasformige luftforureningers effekter på danske plantesamfund. Rapport. Energiministeriet, København.

Johnsen, T.-P. (1997): Quality assurance in analysis and use of environmental data. Nordisk Minister Råd, Copenhagen.

Jonasson, S. (1983): The point intercept method for non-destructive estimation of biomass. Phytocoenologia 11: 385-388.

Jonasson, S. (1988): Evaluation of the point intercept method for the estimation of plant biomass. Oikos 52: 101-106.

Jongejans, E., Schippers, P. (1999): Modeling seed dispersal by wind in herbaceous species. Oikos 87: 362-372.

Jordan, N. (1999): Escape of pest resistance transgenes to agricultural weeds: relevant facets of weed ecology. In: Traynor, P. L., Westwood, J. H. (eds.) Ecological effects of pest resistance genes in managed ecosystems. Information Systems for Biotechnology, http://www.isb.vt.edu, Bethesda, Maryland, USA.

Jordan, N., Mortensen, D.A., Prenzlow, D.M., Cox, K.C. (1995): Simulation analysis of crop-rotation effects on weed seedbanks. Amer. J. Bot. 82: 390-398.

Jørgensen, H.B., Lövei, G.L. (1999): Tri-trophic effect on predator feeding: consumption by the carabid *Harpalus affinis* of *Heliothis armigera* caterpilars fed on proteinase inhibitor-containing diet. Entomol. Exp. Appl. 93: 113-116.

Jørgensen, R.B., Andersen, B. (1994): Spontaneous hybridization between oilseed rape (*Brassica napus* and weedy *B. campestris* (*Brassicaceae*): a risk of growing genetically modified oilseed rape. Amer. J. Bot. 81: 1620-1626.

Jørgensen, R.B., Andersen, B., Hauser, T.P., Landbo, L., Mikkelsen, T.R., Østergard, H., Thomas, G., Monteiro, A.A. (1998): Introgression of crop genes from oilseed rape (*Brassica napus*) to related wild species - An avenue for the escape of engineered genes. Acta Horticult. 459: 211-217.

Jørgensen, R.B., Hauser, T., Mikkelsen, T.R., Østergård, H. (1996): Transfer of engineered genes from crop to wild plants. Trend. Plant Sci. - Perspectives 1: 356-358.

Kapteijns, A.J.A.M. (1993): Risk assessment of genetically modified crops - Potential of 4 arable crops to hybridize with the wild flora. Euphytica 66: 145-149.

Kareiva, P., Parker, I.M., Pascual, M. (1996): Can we use experiments and models in predicting the invasiveness of genetically engineered organisms. Ecology 77: 1670-1675.

Kent, M., Coker, P. (1993): Vegetation description and analysis. John Wiley and Sons, London.

Kjellsson, G., Simonsen, V. (1994): Methods for risk assessment of transgenic plants. I. Competition, establishment and ecosystem effects. Birkhäuser Verlag, Basel.

Kjellsson, G. (1997): Principles and procedures for ecological risk assessment of transgenic plants. In: Kjellsson, G., Simonsen, V., Ammann, K. (eds.) Methods for risk assessment of transgenic plants. II. Pollination, gene-transfer and population impacts. Birkhäuser Verlag, Basel, 221-236.

Kjellsson, G., Simonsen, V., Ammann, K., eds. (1997): Methods for risk assessment of transgenic plants. II. Pollination, gene-transfer and population impacts. Birkhäuser Verlag, Basel.

Kjær, C., Damgaard, C., Kjellsson, G., Strandberg, B., Strandberg, M. (1999): Ecological risk assessment of genetically modified higher plants (GMHP) - Identification of data needs. NERI Technical Report, No 303. National Environmental Research Institute, Silkeborg.

Klein, E.K., Laredo, C. (1999): Optimal sampling designs for studies of gene flow: a comment on Assuncao and Jacobi. Evolution 53: 2002-2005.

Klepper, O., Jager, T., van der Linden, T., Smit, R. (1998): An assessment of the effect on natural vegetations of atmospheric emissions and transport of herbicides in the Netherlands. ECO-memo 98/05. RIVM - Laboratory for Ecotoxicology, Bilthoven.

Klinger, T., Ellstrand, N.C. (1999): Transgene movement via gene flow: recommendations for improved biosafety assessment. In: Ammann, K., Jacot, Y., Simonsen, V., Kjellsson, G. (eds.) Methods for risk assessment of transgenic plants. III. Ecological risks and prospects of transgenic plants, where do we go from here? A dialogue between biotech industry and science. Birkhäuser Verlag, Basel, 129-140.

Kruse, M., Strandberg, M., Strandberg, B. (2000): Ecological effects of allelopathic plants - a review. DMU Technical Report, No. 315. National Environmental Research Institute, Silkeborg.

Landbo, L., Jørgensen, R.B. (1997): Seed germination in weedy *Brassica campestris* and its hybrids with *B. napus*: Implications for risk assessment of transgenic oilseed rape. Euphytica 97: 209-216.

Lavorel, S., Oneill, R.V., Gardner, R.H. (1994): Spatio-temporal dispersal strategies and annual plant species coexistence in a structured landscape. Oikos 71: 75-88.

Lawesson, J.E. (2000): A concept for vegetation studies and monitoring in the Nordic countries. TemaNord 2000: 517. Nordic Council of Ministers, Copenhagen.

Leck, M.A., Parker, V.T., Simpson, R.L. (1989): Ecology of soil seed banks. Academic Press, San Diego.

Lele, S., Taper, M.L., Gage, S. (1998): Statistical analysis of population dynamics in space and time using estimating functions. Ecology 79: 1489-1502.

Levidow, L., Carr, S., Wield, D. (1999): Precautionary commercialization. Safety regulation of transgenic crops: completing the internal market? A study of the implementation of EC Directive 90/220. EC/DGXII/RTD biotechnology programme on the Ethical, Legal and Socio-economic Aspects (ELSA).

Levin, S.A. (1990): Ecological issues related to the release of genetically modified organisms into the environment. In: Mooney, H. A., Bernardi, G. (eds.) SCOPE 44. Introduction of genetically modified organisms into the environment. John Wiley and Sons, Somerset, 151-160.

Lonsdale, W.M. (1999): Global patterns of plant invasions and the concept of invasibility. Ecology 80: 1522-1536.

Lopez-Granados, F., Lutman, P.J.W. (1998): Effect of environmental conditions on the dormancy and germination of volunteer oilseed rape seed (*Brassica napus*). Weed Science 46: 419-423.

Losey, J.E., Rayor, L.S., Carter, M.E. (1999): Transgenic pollen harms monarch larvae. Nature 399: 214.

Mack, R.N. (1989): Temperate grasslands vulnerable to plant invasions: characteristics and consequences. In: Drake, J. A., Mooney, H. A., Castri, F., Groves, R. H., Kruger, F. J., Rejmanek, M., Williamson, M. (eds.) Biological invasions: a global perspective. John Wiley & sons, Chichester, 155-179.

Madsen, K.H., Blacklow, W.M., Jensen, J.E., Streibig, J.C. (1999): Simulation of herbicide use in a crop rotation with transgenic herbicide-tolerant oilseed rape. Weed Research 39: 95-106.

Madsen, K.H., Poulsen, E.R., Streibig, J.C. (1997): Modelling of herbicide use in genetically modified herbicide resistant crops -1. Environmental Project No. 346. Danish Environmental Protection Agency, Copenhagen.

Madsen, K.H., Poulsen, G.S. (1997): Inserted traits for transgenic plants. In: Kjellsson, G., Simonsen, V., Ammann, K. (eds.) Methods for risk assessment of transgenic plants. II. Pollination, gene-transfer and population impacts. Birkhäuser Verlag, Basel, 203-219.

MAFF. (1998): Guide to seed certification in England and Wales. Ministry of Agriculture, Fisheries and Food, London.

Manasse, R.S. (1992): Ecological risks of transgenic plants - Effects of spatial dispersion on gene flow. Ecol. Appl. 2: 431-438.

Mark, S., Strandberg, M. (1999): Modeller til bestemmelse af naturkvalitet på udvalgte naturtyper ved anvendelse af neurale netværk. NERI Technical Report, No. 274. National Environmental Research Institute, Rønde.

Marrs, R.H., Frost, A.J. (1997): A microcosm approach to the detection of the effects of herbicide spray drift in plant communities. J. Environ. Manage. 50: 369-388.

Marrs, R.H., Frost, A.J., Plant, R.A., Lunnis, P. (1993): Determination of buffer zones to protect seedlings of non-target plants from the effects of glyphosate spray drift. Agr. Ecosyst. Environ. 45: 283-293.

Marvier, M.A., Meir, E., Kareiva, P.M. (1999): How do the design of monitoring and control strategies affect the chance of detecting and containing transgenic weeds? In: Ammann, K., Jacot, Y., Simonsen, V., Kjellsson, G. (eds.) Methods for risk assessment of transgenic plants. III. Ecological risks and prospects of transgenic plants, where do we go from here? A dialogue between biotech industry and science. Birkhäuser Verlag, Basel, 109-122.

Mayer, S. (2000): Farm scale trials and environmental safety. In: Allsopp, M., Parr, D. (eds.) GM on trial. Greenpeace, London, 77-83.

Mazyad, R.A. (1997): Morphological character analysis. In: Kjellsson, G., Simonsen, V., Ammann, K. (eds.) Methods for risk assessment of transgenic plants. II. Pollination, gene-transfer and population impacts. Birkhäuser Verlag, Basel, 140-141.

Menges, E.S. (2000): Population viability analyses in plants: challenges and opportunities. Trend. Ecol. Evolut. 15: 51-56.

Metz, P.L.J., Jacobsen, E., Stiekema, W.J. (1997): Aspects of the biosafety of transgenic oilseed rape (*Brassica napus* L). Acta Bot. Neerl. 46: 51-67.

Metz, P.L.J., Jacobsen, E., Stiekema, W.J. (1997): Occasional loss of expression of phosphinothricin tolerance in sexual offspring of transgenic oilseed rape (*Brassica napus* L.). Euphytica 98: 189-196.

Metz, P.L.J., Nap, J.P. (1997): A trans-gene approach to the biosafety of transgenic plants: overview of selection and reporter genes. Acta Bot. Neerl. 46: 25-50.

Mikkelsen, T.R., Andersen, B., Jørgensen, R.B. (1996): The risk of crop transgene spread. Nature 380: 31.

Miljøstyrelsen. (1998): Bekæmpelsesmiddelstatistik 1997. Orientering fra Miljøstyrelsen nr. 6. Miljøstyrelsen, København.

Moldenke, A.R. (1994): Arthropods. In: Weaver, R.W., Angle, S., Bottomley, P., Bezdieck, D., Smith, S., tabatabai, A., Wollum, A. (eds.) Methods of soil analysis. Part 2. Microbiological and biochemical properties. Soil Science Society of America, Madison, Wisconsin, 517-542.

Moyes, C.L., Dale, P.J. (1999): Organic farming and gene transfer from genetically modified crops. John Innes Centre, Norwich.

Nielsen, J., Strandberg, M., Tybirk, K., Groth, N.B. (1998): Det økologiske råderum for areal og fødevarer - et definitions- og metodeprojekt. Arbejdsrapport fra DMU, Nr. 92. DMU, Roskilde.

Norris, C.E., Simpson, E.C., Sweet, J.B., Thomas, J.E. (1999): Monitoring weediness and persistence of genetically modified oilseed rape (*Brassica napus*) in the UK. In: Lutman, P. J. W. (ed.) BCPC Symposium Proceedings No. 72. Gene flow and agriculture - Relevance for transgenic crops, British Crop Protection Council, University of Keele, Staffordshire, 255-260.

Nygaard, B., Lawesson, J., Ejrnæs, R. (1999a): DANVEG - computersoftware. National Environmental Research Institute, Rønde.

Nygaard, B., Mark, S., Baattrup-Pedersen, A., Dahl, K., Ejrnæs, R., Fredshavn, J., Hansen, J., Lawesson, J., Münier, B., Møller, P.F., Risager, M., Rune, F., Skriver, J., Søndergaard, M. (1999b) Naturkvalitet - kriterier og metodeudvikling, Technical report 285, National Environmental Research Institute, Rønde.

Odderskær, P., Prang, A., et al. (1997): Skylark reproduction in pesticide treated and untreated fields. Ministry of the Environment and Energy, Danish Environmental Protection Agency, Copenhagen.

OECD. (1993): Safety considerations for biotechnology: scale-up of crop plants. OECD, Paris.

OECD. (1997a): Consensus document on the biology of *Brassica napus* L. (Oilseed Rape). Series on Harmonization of Regulatory Oversight in Biotechnology, No. 7. OECD, Paris.

OECD. (1997b): Consensus document on the biology of *Solanum tuberosum* subsp. *tuberosum* (Potato). Series on Harmonization of Regulatory Oversight in Biotechnology, No. 8. OECD, Paris.

OECD. (1999a): Consensus document on the biology of *Triticum aestivum* (bread wheat). Series on Harmonization of Regulatory Oversight in Biotechnology, No. 9. OECD, Paris.

OECD. (1999b): Consensus document on the biology of *Picea abies*(L.) Karst (Norway spruce). Series on Harmonization of Regulatory Oversight in Biotechnology, No. 12. OECD, Paris.

Oppenheimer, Wolff, Donnelly. (1996): Elaboration on possible arguments and objectives of an additional EC policy on plant protection products. E.C., Brussels.

Parker, I.M., Bartsch, D. (1996): Recent advances in ecological biosafety research on the risks of transgenic plants: A trans-continental perspective. In: Tomiuk, J., Wöhrmann, K., Sentker, A. (eds.) Transgenic organisms: Biological and social implications. Birkhäuser Verlag, Basel, 147-161.

Parker, I.M., Kareiva, P. (1996): Assessing the risks of invasion for genetically engineered plants: acceptable evidence and reasonable doubt. Biol. Conserv. 78: 193-203.

Parker, P.G., Snow, A.A., Schug, M.D., Booton, G.C., Fuerst, P.A. (1998): What molecules can tell us about populations: choosing and using a molecular marker. Ecology 79: 361-382.

Pekrun, C., Lane, P.W., Lutman, P.J.W. (1999): Modelling the potential for gene escape in oilseed rape via the soil seedbank: its relevance for genetically modified cultivars. In: Lutman, P. J. W. (ed.) BCPC Symposium Proceedings No. 72. Gene flow and agriculture - Relevance for transgenic crops. British Crop Protection Council, University of Keele, Staffordshire, 101-106.

Petersen, H., Luxton, M. (1982): A comparative analysis of soil fauna populations and their role in decomposition. Oikos 39: 287-388.

Pfeilstetter, E., Matzk, A., Feldmann, S.D., Schiemann, J. (2000): Rapid and efficient screening of phosphinothricin tolerant oilseed rape *(Brassica napus)* with a novel germination test. Euphytica 113: 119-124.

Philippi, T.E., Dixon, P.M., Taylor, B.E. (1998): Detecting trends in species composition. Ecol. Appl. 8: 300-308.

Pohl-Orf, M., Brand, U., Driessen, S., Hesse, P.R., Lehnen, M., Morak, C., Mucher, T., Saeglitz, C., von Soosten, C., Bartsch, D. (1999): Overwintering of genetically modified sugar beet, *Beta vulgaris* L. subsp *vulgaris*, as a source for dispersal of transgenic pollen. Euphytica 108: 181-186.

Powell, T.M., Steele, M., eds. (1995): Ecological time series. Chapman & Hall, New York.

Price, J.S., Hobson, R.N., Neale, M.A., Bruce, D.M. (1996): Seed losses in commercial harvesting of oilseed rape. J. Agr. Eng. Res. 65: 183-191.

Proctor, M., Yeo, P., Lack, A. (1996): The natural history of pollination. Timber Press, Portland.

RAFI. (2000): Terminator two years later: RAFI update on terminator/traitor technology. The Rural Advancement Foundation International, Canada.

Rao, J.N.K., Graham, J.E. (1964): Rotation designs for sampling on repeated occasions. J. Am. Stat. Ass. 59: 492-509.

Reddersen, J. (1997): The arthropod fauna of organic versus conventional cereal fields in Denmark. Biol. Agri. Hort. 15: 201-211.

Regal, P.J. (1993): The true meaning of "exotic species" as a model for genetically engineered organisms. Experientia 49: 225-234.

Rejmanek, M. (1989): Invasibility of plant communities. In: Drake, J. A., Mooney, H. A., Castri, F., Groves, R. H., Kruger, F. J., Rejmanek, M., Williamson, M. (eds.) Biological invasions: a global perspective. John Wiley & sons, Chichester, 369-388.

Rew, L.J., Cussans, G.W. (1997): Horizontal movement of seeds following tine and plough cultivation: implications for spatial dynamics of weed infestations. Weed Research 37: 247-256.

Reynolds, J.H., ed. (1998a): WCMC Handbooks on biodiversity information management, Vol. 1-7. World Conservation Monitoring Centre, Commonwealth Secretariat, London.

Reynolds, J.H., ed. (1998b): Information and policy. WCMC Handbooks on biodiversity information management, Vol. 1. World Conservation Monitoring Centre, Commonwealth Secretariat, London.

Reynolds, J.H., ed. (1998c): Information networks. WCMC Handbooks on biodiversity information management, Vol. 4. World Conservation Monitoring Centre, Commonwealth Secretariat, London.

Risager, M. (1998): Effect of nitrogen deposition on Sphagnum dominated bogs. Ph.D. thesis. Botanical Institute, Copenhagen.

Rissler, J., Mellon, M. (1993): Perils among the promise: ecological risks of transgenic crops in a global market. Union of Concerned Scientists, Cambridge.

Romao, C. (1996): Interpretation manual of European Union habitats, EUR 15. European Commission, DG XI, Bruxelle.

Ruckelshaus, M., Hartway, C., Kareiva, P. (1997): Assessing the data requirements of spatially explicit dispersal models. Conserv. Biol. 11: 1298-1306.

Rufener Al Mazyad, P., Ammann, K. (1999): Biogeographical assay and natural gene flow. In: Ammann, K., Jacot, Y., Simonsen, V., Kjellsson, G. (eds.) Methods for risk assessment of transgenic plants III. Ecological risks and prospects of transgenic plants, where do we go from here? A dialogue between biotech industry and science. Birkhäuser Verlag, Basel, 95-98.

Ruffio-Chable, V., Bellis, H., Herve, Y. (1993): A dominant gene for male-sterility in cauliflower (*Brassica oleracea* var *botrytis*): phenotype expression, inheritance, and use in F1-hybrid production. Euphytica 67: 9-17.

Röttgermann, M., Steinlein, T., Beyschlag, W., Dietz, H. (2000): Linear relationships between aboveground biomass and plant cover in low open herbaceous vegetation. J. Veg. Sci. 11: 145-148.

Raamsdonk, L.W.D.v., van der Maesen, L.J.G. (1996): Crop-weed complexes: the complex relationship between crop plants and their wild relatives. Acta Bot. Neerl. 45: 135-155.

Scheffler, J.A., Parkinson, R., Dale, P.J. (1993): Frequency and distance of pollen dispersal from transgenic oilseed rape (*Brassica napus*). Transgenic Research 2: 57-65.

Scheffler, J.A., Parkinson, R., Dale, P.J. (1995): Evaluating the effectiveness of isolation distances for field plots of oilseed rape (*Brassica napus*) using a herbicide-resistance transgene as a selectable marker. Plant Breed. 114: 317-321.

Schuphan, I. (1999): Results from biosafety rersearch with transgenic sugar beet: ecological impact of virus resistance in cropland and wild beet habitats. In: de Vries, G. E. (ed.) Past, present and future considerations in risk assessment when using GMO's. Commission Genetic Modification, Bilthoven, 89-98.

Scott, S.E., Wilkinson, M.J. (1998): Transgene risk is low. Nature 393: 320.

Senior, I.J., Dale, P.J. (1999): Molecular aspects of multiple transgenes and gene flow to crops and wild relatives. In: Lutman, P. J. W. (ed.) BCPC Symposium Proceedings No. 72. Gene flow and agriculture - Relevance for transgenic crops. British Crop Protection Council, University of Keele, Staffordshire, 225-232.

Shaffers, A.P., Sýkora, K.V. (2000): Reliability of Ellenberg indicator values for moisture, nitrogen and soil reaction: a comparison with field measurements. J. Veg. Sci. 11: 225-244.

Simonsen, V. (1999): Molecular markers for monitoring transgenic plants. In: Ammann, K., Jacot, Y., Simonsen, V., Kjellsson, G. (eds.) Methods for risk assessment of transgenic plants III. Ecological risks and prospects of transgenic plants, where do we go from here? A dialogue between biotech industry and science. Birkhäuser Verlag, Basel, 87-93.

Skalski, J.R. (1990): A design for long-term monitoring. J. Environ. Manage. 30: 139-144.

Snow, A.A., Jørgensen, R.B. (1999): Fitness costs associated with transgenic glufosinate tolerance introgressed from *Brassica napus* ssp. *oleifera* (oilseed rape) into weedy *Brassica rapa*. In: Lutman,

P. J. W. (ed.) BCPC Symposium Proceedings No. 72. Gene flow and agriculture - Relevance for transgenic crops. British Crop Protection Council, University of Keele, Staffordshire, 137-142.

Snow, A.A., Palma, P.M. (1997): Commercialization of transgenic plants: potential ecological risks. Bioscience 47: 86-96.

Sokal, R.R., Rohlf, F.J. (1995): Biometry - The principles and practice of statistics in biological research. W.H. Freeman, New York.

Southwood, T.R.E. (1978): Ecological methods with particular reference to the study of insect populations. Chapman and Hall, London.

Spellerberg, I.F. (1991): Monitoring ecological change. Cambridge University Press, Canbridge.

Steward, C.N., All, J.N., Raymer, P.L., Ramachandran, S. (1997): Increased fitness of transgenic insecticidal rapeseed under insect selection pressure. Mol. Ecol. 6: 773 - 779.

Stewart, C.A., Black, V.J., Black, C.R., Roberts, J.A. (1996): Direct effects of ozone on the reproductive development of *Brassica* species. J. Plant Physiol. 148: 172-178.

Stewart, C.N. (1996): Monitoring transgenic plants using in vivo markers. Nature Biotechnology 14: 682.

Stewart, C.N.J. (1999): Insecticidal transgenes into nature: gene flow, ecological effects, relevancy and monitoring. In: Lutman, P. J. W. (ed.) BCPC Symposium Proceedings No. 72. Gene flow and agriculture - Relevance for transgenic crops. British Crop Protection Council, University of Keele, Staffordshire, 179-190.

Stewart, J., Potvin, C. (1996): Effects of elevated CO^2 on an artificial grassland community - Competition, invasion and neighbourhood growth. Funct. Ecol. 10: 157-166.

Stohlgren, T.J., Binkley, D., Chong, G.W., Kalkhan, M.A., Schell, L.D., Bull, K.A., Otsuki, Y., Newman, G., Bashkin, M., Son, Y. (1999): Exotic plant species invade hot spots of native plant diversity. Ecol. Monogr. 69: 25-46.

Strandberg, B., Kjellsson, G., Løkke, H. (1998): Hierarchial risk assessment of transgenic plants: Proposal for a new system. Biosafety 4, http://www.btd.org.br/bioline.

Strykstra, R.J., Verweij, G.L., Bakker, J.P. (1997): Seed dispersal by mowing machinery in a Dutch brook valley system. Acta Bot. Neerl. 46: 387-401.

Stute, K. (1991): Richtlinien für die Prüfung von Pflanzenschutzmitteln in Zulassungsverfahren Teil VI., Biologische Bundesanstalt für Land- und Forstwirtschaft Bundesrepublik Deutschland, Braunschweig.

Sweet, J.B., Shepperson, R., Thomas, J.E., Simpson, E. (1997): The impact of releases of genetically modified herbicide tolerant oilseed rape in the UK. 1997 Brighton crop protection conference: weeds, No. 1. Proceedings of an international conference, Brighton, UK, 17-20 November 1997. British Crop Protection Council, Staffordshire, 291-302.

ter Braak, C.J.F. (1996): Unimodal models to relate species to environment. DLO-Agricultural Mathematics Group, Wageningen.

Thomas, L. (1996): Monitoring long-term population change: why are there so many analysis methods? Ecology 77: 49-58.

Thomas, L. (1997): Retrospective power analysis. Conserv. Biol. 11: 276-280.

Thomas, L., Krebs, C.J. (1996): A review of statistical power analysis software. Bull. Ecol. Soc. Amer. 78: 128-139.

Thompson, K., Bakker, J.P., Bekker, R.M. (1997): The soil seed banks of North West Europe: methodology, density and longevity. Cambridge University Press, Cambridge.

Thompson, K., Hodgson, J.G., Grime, J.P., Rorison, I.H., Band, S.R., Spencer, R.E. (1993): Ellenberg numbers revisited. Phytocoenologia 23: 277-289.

Thompson, K., Hodgson, J.G., Rich, T.C.G. (1995): Native and alien invasive plants: more of the same? Ecography 18: 390-402.

Tiedje, J.M., Colwell, R.K., Grossman, Y.L., Hodson, R.E., Lenski, R.E., Mack, R.N., Regal, P.J. (1989): The planned introduction of genetically engineered organisms: Ecological considerations and recommendations. Ecology 70: 298-315.

Tilman, D. (1997): Community invasibility, recruitment limitation, and grassland biodiversity. Ecology 78: 81-92.

Tilman, D., Downing, J.A. (1994): Biodiversity and stability in grasslands. Nature 367: 363-365.

Trevors, J.T. (1998): Bacterial biodiversity in soil with an emphasis on chemically-contaminated soils. Water Air Soil Pollut. 101: 45-67.

Tybirk, K., Strandberg, B. (1999): Oak forest development as a result of historical land-use patterns and present nitrogen deposition. For. Ecol. Man. 114: 97-106.

Tyler, T., Olsson, K.-A. (1997): Förändringar i Skånes flora under perioden 1938-1996 - statistisk analys av resultat från två inventeringar. Svensk Bot. Tidskr. 91: 143-185.

UBA. (1996): Manual on methodologies and criterias for mapping critical levels/loads and geographical areas where they are exceeded. Texte UmweltsBundesamt 71/96. UBA, Berlin.

Underwood, A.J. (1994): On beyond BACI: sampling designs that might reliably detect environmental disturbances. Ecol. Appl. 4: 3-15.

UN-ECE. (1998): Manual for integrated monitoring., Finnish Environment Institute, Helsinki.

UNEP. (1996): International technical guidelines for safety in biotechnology. United Nations Environmental Programme, Nairobi.

Urquhart, N.S., Paulsen, S.G., Larsen, D.P. (1998): Monitoring for policy-relevant regional trends over time. Ecol. Appl. 8: 246-257.

van Hees, W.W.S., Mead, B.R. (2000): Ocular estimates of understorey vegetation structure in a closed *Picea glauca / Betula papyrifera* forest. J. Veg. Sci. 11: 195-200.

Vos, P., Meelis, E., ter Keurs, W.J. (2000): A framework for the design of ecological monitoring programs as a tool for environmental and nature management. Environ. Monit. Assess. 61: 317-344.

Waddington, K.D. (1983): Foraging behaviour of pollinators. In: Real, L. (ed.) Pollination Biology, Academic Press, Orlando, 213-239.

Weaver, R.W., Angle, S., Bottomley, P., Bezdieck, D., Smith, S., tabatabai, A., Wollum, A., eds. (1994): Methods of soil analysis. Part 2. Microbiological and biochemical properties. Soil Science Society of America, Madison, Wisconsin.

Welles, J.M. (1990): Some indirect methods for estimating canopy structure. Remote Sens. Rev. 5: 31-43.

Wijnheimer, E.H.M., Brandenburg, W.A., Ter Borg, S.J. (1989): Interactions between wild and cultivated carrots (*Daucus carota* L.) in the Netherlands. Euphytica 40: 147-154.

Wilkinson, J.E., Twell, D., Lindsey, K. (1997): Activities of Camv 35s and Nos promoters in pollen: implications for field release of transgenic plants. J. Exp. Bot. 48: 265-275.

Wilkinson, M.J., Timmons, A.M., Charters, Y., Dubbels, S., Robertson, A., Wilson, N., Scott, S., O'Brian, E., Lawson, H.M. (1995): Problems of risk assessment with genetically modified oilseed rape. Brighton Crop Protection Conference - Weeds 1995. British Crop Protection Council, Staffordshire, 1035-1044.

Williamson, M. (1993): Invaders, weeds and the risk from genetically manipulated organisms. Experientia 49: 219-224.

Williamson, M. (1999): Invasions. Ecography 22: 5-12.

Wilson, J.B., Gitay, H., Steel, J.B., King, W.M. (1998): Relative abundance distributions in plant communities: effects of species richness and of spatial scale. J. Veg. Sci. 9: 213-220.

Wilson, S.D., Tilman, D. (1991): Components of plant competition along an experimental gradient of nitrogen availability. Ecology 72: 1050-1065.

Wollum, A.G.I. (1994): Soil sampling for microbiological analysis. In: Weaver, R. W., Angle, S., Bottomley, P., Bezdieck, D., Smith, S., tabatabai, A., Wollum, A. (eds.) Methods of soil analysis. Part 2. Microbiological and biochemical properties. Soil Science Society of America, Madison, Wisconsin, 1-14.

Young, R. (2000): Risks to organic farming. In: Allsopp, M., Parr, D. (eds.) GM on trial. Greenpeace, London, 19-30.

11 Subject index